Conference Proceedings of the Society for Experimental Mechanics Series

Series Editor

Kristin B. Zimmerman
Society for Experimental Mechanics, Inc.,
Bethel, USA

The Conference Proceedings of the Society for Experimental Mechanics Series presents early findings and case studies from a wide range of fundamental and applied work across the broad range of fields that comprise Experimental Mechanics. Series volumes follow the principle tracks or focus topics featured in each of the Society's two annual conferences: IMAC, A Conference and Exposition on Structural Dynamics, and the Society's Annual Conference & Exposition and will address critical areas of interest to researchers and design engineers working in all areas of Structural Dynamics, Solid Mechanics and Materials Research.

Paul Allison • Phillip Jannotti
Editors

Dynamic Behavior of Materials, Vol. 1

Proceedings of the 2025 Annual Conference
on Experimental and Applied Mechanics

River Publishers

Published, sold and distributed by:

River Publishers
Broagervej 10
9260 Gistrup
Denmark

www.riverpublishers.com

ISBN 97887-438-0827-5 (Hardback)

ISBN 97887-438-0832-9 (eBook)

https://doi.org/10.13052/9788743808275

Conference Proceedings of the Society for Experimental Mechanics

An imprint of River Publishers

Preface

Dynamic Behavior of Materials represents one of five volumes of technical papers presented at the 2025 SEM Annual Conference & Exposition on Experimental and Applied Mechanics organized by the Society for Experimental Mechanics and held in Milwaukee, WI, June 2-5, 2025. The complete proceedings also include volumes on: *Advancement of Optical Methods & Digital Image Correlation in Experimental Mechanics; Mechanics of Biological Systems and Materials and the Mechanics of Composite, Hybrid & Multifunctional Materials; Fracture, Fatigue, Failure, Damage Evolution and Thermomechanics & Infrared Imaging; and Mechanics of Additive & Advanced Manufacturing, Inverse Methods and Machine Learning.*

Each collection presents early findings from experimental and computational investigations on an important area within Experimental Mechanics. Dynamic Behavior of Materials is one of these areas.

The Dynamic Behavior of Materials track was initiated in 2005 and reflects our efforts to bring together researchers interested in the dynamic behavior of materials and structures, and to provide a forum to facilitate technical interaction and exchange. Over the years, this track has been representing the ever-growing interests in dynamic behavior to the SEM community, working towards expanding synergy with other tracks and topics, and improving diversity and inclusivity, as evidenced by the increasing number and diversity of papers and attendance.

The contributed papers span numerous technical divisions within SEM, demonstrating its relevance not only in the dynamic behavior of materials community, but also in the mechanics of materials community as a whole. The track organizers thank the authors, presenters, organizers and session chairs for their participation, support, and contribution to this track. The SEM support staff is also acknowledged for their devoted efforts in accommodating the large number of paper submissions this year, making the 2025 Dynamic Behavior of Materials Track a success.

Editors:
Paul Allison – *Baylor University, Tx, USA;* Phillip Jannotti—*U.S. Army Research Laboratory, MD, USA*

Contents

Chapter 1
The Influence of Entrained Air on the Dynamic Compressive Performance of Conventional Concrete

Mahmoud M. Awd Allah*, **Mohamed A. Abbas, Brett A. Williams, William F. Heard, and Xu Nie**

Abstract Freeze-thaw damage is a common form of deterioration in concrete that happens when it is subjected to frequent cycles of freezing and thawing, mainly in cold weather. This kind of damage mostly affects the durability of concrete, causing scaling, cracking, and spalling that over time can compromise the structural integrity. One of the useful approaches to decreasing freeze-thaw damage is by adjusting permeability, such as introducing entrained air in the concrete mix acting as pressure-relief zones when water inside the concrete freezes and expands. This method does not directly reduce permeability but mitigates the influence of any water entrapped in the concrete. Understanding the relationship between the entrained air percent and dynamic behaviors of concrete facilitates more effective use of materials. By studying this connection, designers can determine the acceptable percentage of air that doesn't significantly compromise the concrete's performance. It is therefore the goal of this study to answer the question: To what extent can air content affect the dynamic compressive performance of conventional concrete? Subsequently, cylindrical specimens with different air content percentages were developed and tested under dynamic strain rates. For the dynamic testing, a 2-inch diameter Kolsky compression bar system with well-designed pulse shapers was used to enable the constant strain rate deformation and dynamic stress equilibrium conditions. The results revealed that the air content imposes significant impacts on the dynamic compressive performance of conventional concrete.

Keywords Heterogeneous material · Freeze-thaw damage · Permeability · High strain rate · Split Hopkinson pressure bar

Introduction

Concrete is the most widely used building material due to its cost-effectiveness, durability, mechanical properties, and design flexibility [1]. However, it faces challenges that affect its performance and sustainability, particularly freeze-thaw (FT) damage in regions with fluctuating temperatures around freezing point [2]. During freezing, internal moisture in concrete expands, creating internal pressures that can cause micro cracks and damage. In the thaw cycle, the ice melts, and the concrete contracts, further stressing the structure [3]. Repeated FT cycles worsen the damage, weakening key properties like mass, elasticity, strength, and fracture toughness.

Permeability is a key factor in FT damage, especially in materials like concrete and soil. It refers to a material's ability to allow fluids, such as water, to pass through [4]. High permeability allows fluids to pass through easily, while low permeability means fluids flow slowly or are blocked. This property is critical in fields like geology, civil engineering, and environmental science, as it influences water movement, drainage, and structural stability. To address FT damage in concrete related to permeability, a variety of solutions have been developed and implemented to improve its resistance and durability. One of the solutions is design the concrete mix by incorporating air-entraining agents into the mix [5]. These agents create microscopic air bubbles that allow for the expansion of freezing water without causing damage. For this instant, according to Peng et al. [6] the inclusion of entrained air significantly enhances frost resistance, allowing materials to withstand up to 300 FT cycles. This improvement is attributed to the air bubbles created during the mixing process, which act as a cushion, absorbing the stress generated by the expansion of freezing water within the material's pores.

Mahmoud M. Awd Allah · Mohamed A. Abbas · Xu Nie
Mechanical Engineering Department, Southern Methodist University, 6425 Boaz Lane, Dallas, TX 75205, USA
e-mail: mawdallah@smu.edu; maabbas@mail.smu.edu; xnie@mail.smu.edu

Brett A. Williams · William F. Heard
Army Engineer Research and Development Center, 3909 Halls Ferry Rd, Vicksburg, MS 39180, USA
e-mail: brett.a.williams@erdc.dren.mil; william.f.heard@erdc.dren.mil

Paul Allison et al. (eds.), *Dynamic Behavior of Materials, Vol. 1*, Conference Proceedings of the Society for Experimental Mechanics Series, https://doi.org/10.13052/9788743808275-1

Additionally, Tanesi and Meininger [7] investigated the FT resistance of concrete with minimal air content by utilizing Vinsol resin and synthetic air-entraining agents. The findings revealed that concrete with an air content of up to 3.5% exhibited satisfactory FT performance when combined with Vinsol resin-based air-entraining agents. This suggests that the specific properties of Vinsol resin contribute positively to the air-entraining process, enhancing the material's resistance to FT cycles. Later on, Hang et al. [8] pointed out that concrete displays the best FT resistance at air content of 4.6 : 6.5%. Moreover, the addition of entrained air affects the concrete's fluidity and compressive strength. The concrete's fluidity first rises and then falls with an increase in entrained air dosage, and the compressive strength decrease. It is also noteworthy that when comparing concrete created using recycled aggregate to concrete made with virgin aggregate, the use of air entrainment and polypropylene fibers has proven to be similarly successful in providing FT endurance, as stated by Richardson et al. [9].

Considering the literature listed above, there is a lack of experimental studies on the dynamic compressive performance of conventional concrete with entrained air. To address this, cylindrical specimens with air content percentages of 2 and 7% were developed and tested under dynamic strain rates. A 2-inch diameter Kolsky compression bar system, equipped with pulse shapers, was used for the dynamic testing to maintain constant strain rate deformation and dynamic stress equilibrium conditions.

Methodology

Materials

The material being utilized in this study is conventional concrete, that has been developed by the U.S. Army Engineer Research and Development Center (ERDC). Conventional concrete, known for its strength, versatility, and durability, is formulated by combining specific proportions of various ingredients such as cement, aggregates, water, and admixtures. Table 1 provides a detailed breakdown of the mixture proportions for the individual components of this particular concrete.

Specimens preparation

A four-step process described by Heard et al. [10] was used to prepare all specimens used in this work. Once the material is fully cured, the process begins by core drilling long cylinders with 44.60-mm in diameter from the bulk material. These cylinders are then sectioned into individual pieces using a Beuhler Delta-EB 260 manual abrasive cutter. Subsequently, the ends of the specimens undergo grinding using a Harig 618 auto-step precision surface-grinder to enhance the flatness and parallelism of the ends. Finally, each specimen undergoes polishing in a high-precision P.R. Hoffman PR-1 double-sided lapping machine to achieve their final length of 22.40 mm. This particular process ensures maximum flatness and parallelism of the specimen ends, thereby minimizing stress concentrations at the specimen/bar interfaces.

Table 1 The mixture proportions for the concrete used in this work.

Component	Unit	Amount (per one cubic yard)
Type 1L Cement	lbs	517.5
#89 Limestone	lbs	1558.7 (absorption: 0.30%)
Concrete Sand	lbs	1496.3 (absorption: 0.50%)
Water	lbs	290.9
Admixture 1 – Viscocrete 2100	lbs	2.19
Admixture 2 – Sikatard	lbs	2.91
Admixture 3 – Sika Air	lbs	0.20
Water to cement ratio (w/c)	-	0.57

Test description

The dynamic uniaxial compression tests were conducted using a 2-inches diameter Kolsky compression bar apparatus. The Kolsky Compression Bar, also referred to as the Split-Hopkinson pressure bar, is a dependable method for evaluating material response under high strain-rate uniaxial compression. The method is based on one dimensional stress wave theory in elastic solids. In this test setup, the specimen is placed between two elongated elastic rods, forming the conventional Kolsky compression bar assembly, aligned along a shared axis. A gas gun typically propels a striker, accelerating it to collide with the incident bar, which is one end of the first rod. The impact generates a compressive stress wave that travels along the incident bar towards the specimen. Upon reaching the specimen end, a portion of this stress wave reflects back up the incident bar, while the rest transmits into the specimen. This transmission occurs due to a shift in mechanical impedance at the specimen interface. The specimen is deformed when the transmitted stress wave passes through it and enters the second rod, also referred to as the transmission bar. Resistive strain gauges attached to the bar surfaces allow for the measurement and recording of the stress waves as they move along the bars. A Wheatstone bridge circuit is coupled to the strain gauges, and a high-resolution digital oscilloscope records the output voltage. The oscilloscope's recorded waves can be used to compute the specimen's stress and deformation history if the stress waves are assumed to be one-dimensional and non-dispersive. A well-designed experiment can satisfy these two essential assumptions. It has been demonstrated that the one-dimensional wave propagation assumption holds true provided that the incident and transmission bars are long enough to enable the radial stress distribution to become uniform. Wave dispersion effects can become significant when the incident loading wave comprises components of different frequencies. However, studies have indicated that employing proper pulse shaping techniques to eliminate higher frequency components of the incident loading wave can render the effects of wave dispersion negligible [11]. Figure 1 represents a schematic diagram of the Kolsky compression bar used in the present work.

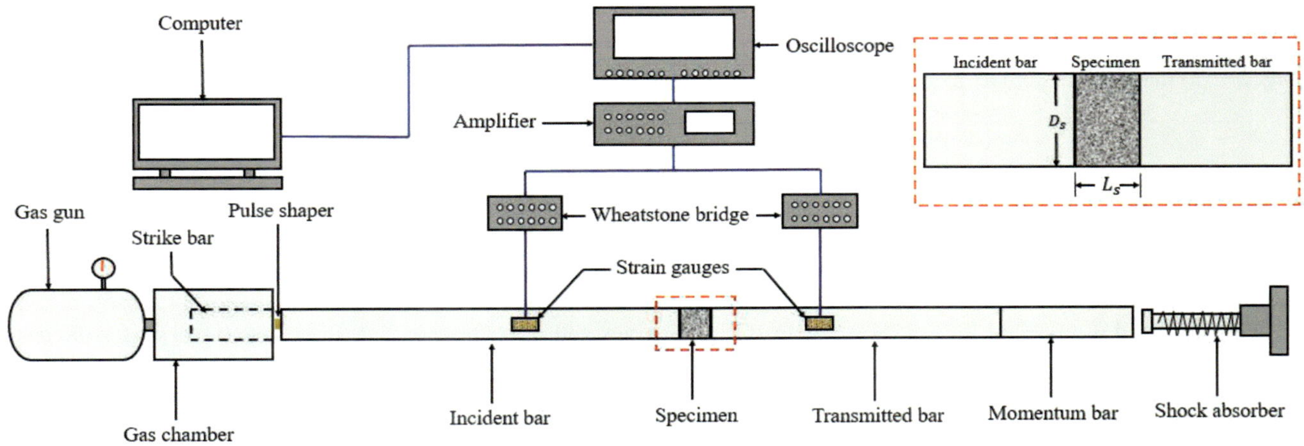

Fig. 1 Schematic diagram of Kolsky bar.

Results and Discussions

To evaluate the average dynamic compressive strength of the studied material with different air percentages, five samples were tested using a Kolsky bar apparatus under monotonic compressive loading conditions. The dynamic compressive strength for each specimen was determined by identifying the maximum stress recorded during the loading phase. This maximum stress represents the material's peak resistance to dynamic compression prior to fragmentation. Figure 2 illustrates the original waveforms obtained from the oscilloscope during the monotonic loading tests, including the incident, reflected, and transmitted wave components. Additionally, Figure 2 presents the calculated bar stress histories at two critical locations: the incident bar end ($\sigma_I + \sigma_R$), where the sum of incident and reflected stresses is analyzed, and the transmission bar end (σ_T), where the transmitted stress is recorded.

The strain rate of the specimen is directly proportional to the behavior of the reflected wave during the loading process. This relationship allows the strain rate history during monotonic loading to be categorized into three distinct stages, each representing a specific phase of deformation and failure. The first stage is characterized by a progressive increase in the amplitude of the reflected wave. This increase corresponds to a period of non-constant strain rate deformation, which arises due to the axial acceleration of the specimen particles. During this phase, the material is transitioning from a state of rest to dynamic loading, and the strain rate varies as the specimen begins to resist the applied load. In the second stage, the reflected wave profile reaches a plateau, indicating the establishment of a constant strain rate. The plateau demonstrates a stable deformation response, where the applied stress and resulting strain rate remain balanced [12]. This stage is critical for accurately capturing the dynamic behavior of the material under controlled loading conditions. The third stage is marked by a sharp and dramatic increase in the amplitude of the reflected wave. At this point, the wave no longer represents the strain rate of the specimen. Instead, it reflects the free motion of the bar's end face, which occurs as the specimen fails and

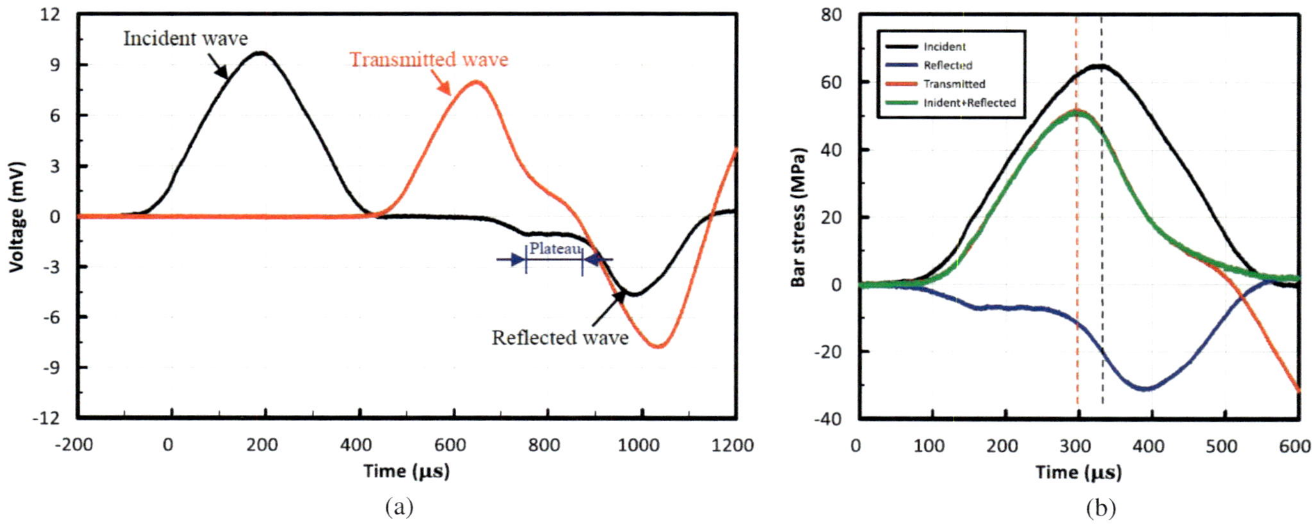

Fig. 2 (a) the original signals recorded by strain gauges and (b) the shifted waves.

can no longer sustain the applied load. This phase signals the peak of the loading process and the complete fragmentation of the specimen. Additionally, the transmitted wave provides further insight into the material's response. The transmitted wave reaches its peak amplitude earlier than the incident wave, a phenomenon attributed to the material's inherent softening behavior. The trailing portion of the transmitted wave, results from this softening effect and highlights the gradual reduction in the material's ability to carry load as failure progresses.

Figures 3 and 4 represent the engineering stress-strain curves and the corresponding strain rate histories for the tested specimens with 2 and 7% under monotonic compression loading, respectively. It is clear that both the percentages demonstrate an initial elastic response, which transitions into plastic strain hardening as the material deforms toward its peak stress. The transition from elastic to plastic behavior is primarily attributed to the propagation and combination of micro cracks. This mechanism mitigates the high lateral tensile stress concentrations that develop at the crack tips under compressive loading conditions. These micro cracks play a vital role in accommodating the compressive forces and contribute to the overall deformation characteristics of the material. The heterogeneous composition of the concrete material introduces some variability in the dynamic mechanical response observed across individual specimens. This variability arises from the inherent differences in the material's microstructure, such as aggregate distribution and pore size, which influence its strength and deformation behavior under dynamic loading. Despite the inherent variability, the experimental data shown in Figures 3 and 4 demonstrate a notable reduction in scatter within the stress-strain responses. This enhanced consistency can largely be attributed to the implementation of pulse shaping techniques. These techniques are essential in ensuring that the loading conditions remain both controlled and repeatable [13]. By effectively smoothing the incident wave, pulse shaping minimizes fluctuations and irregularities in the applied load, promoting a more uniform distribution of stress across the specimen. As a result, these techniques help to reduce experimental artifacts, leading to more reliable, consistent, and reproducible outcomes that strengthen the overall validity of the experimental findings.

Additionally, Table 2 demonstrates the average mechanical properties of concrete specimens with varying air content percentages. The data clearly indicates that the dynamic compressive performance of the specimens with 2% air content is significantly higher, showing a 19.63% increase compared to those with 7% air content. This difference in performance suggests that the air content plays a significant role in the mechanical behavior of concrete under dynamic loading conditions. The improved performance of the specimens with lower air content is the reduction in porosity and air voids within the concrete matrix. Concrete with higher air content typically has an increased number of microscopic voids, which can act as points of weakness under compressive stress. These voids can reduce the overall density and structural integrity of the material, making it more prone to deformation and failure when subjected to dynamic loads, such as impact or high strain rate compression. In contrast, reducing the air content in the concrete leads to a denser material with fewer voids, which enhances its ability to distribute applied stresses more uniformly throughout the matrix. This results in a stronger, more resilient material that is better equipped to withstand dynamic forces. Additionally, lower air content often leads to improved bonding between the cement paste and aggregates, further strengthening the concrete's overall mechanical properties. Furthermore, the improved performance of the 2% air content specimens could be the reduction in the potential for FT damage. Concrete with excessive air content may exhibit greater susceptibility to FT cycles, especially in cold climates. Although air entrain-

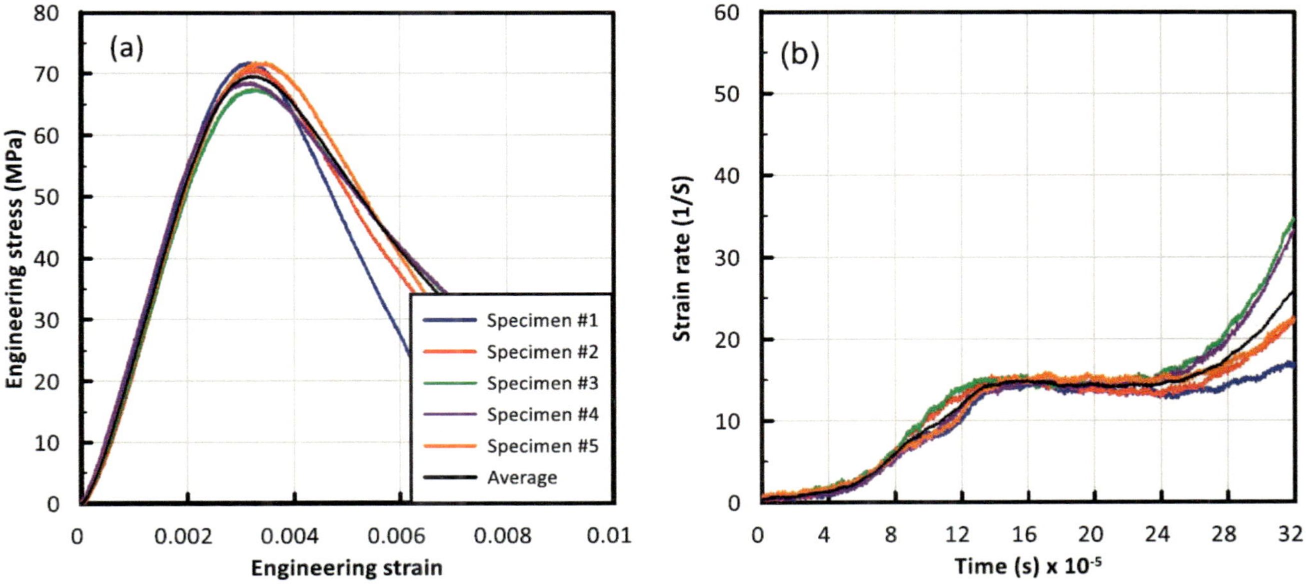

Fig. 3 Dynamic performance (a) compressive stress-strain and (b) strain rate history for specimens with 2% air percent.

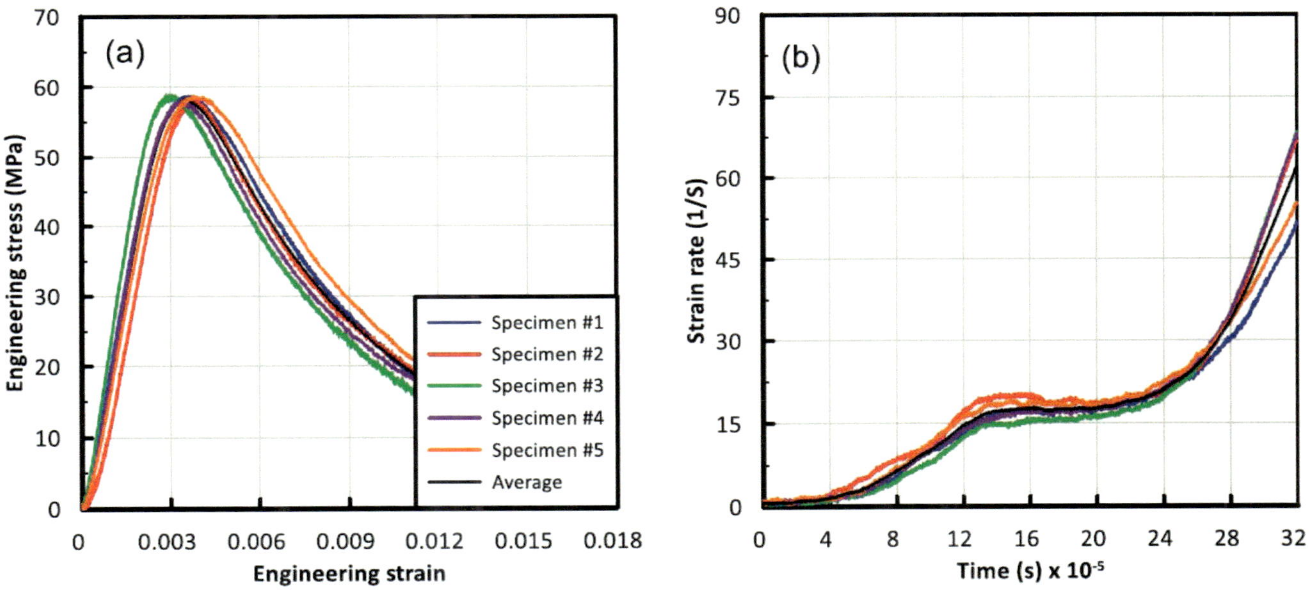

Fig. 4 Dynamic performance (a) compressive stress-strain and (b) strain rate history for specimens with 7% air percent.

ment is typically used to improve the FT resistance of concrete, an optimal balance must be maintained. Too much air can lead to excessive void spaces that, while initially providing some resistance to freezing and thawing, could also reduce the concrete's overall strength and performance in dynamic conditions. These findings emphasize the importance of controlling air content in concrete mixtures to optimize their mechanical properties, particularly for applications that require resistance to dynamic loads and compressive forces. This result is consistent with the findings of Ge et al. [14], who reported that properly managing air content ensures that the concrete achieves a balance between durability, strength, and performance, leading to more reliable results in practical engineering applications.

Moreover, Table 2 displays the plastic strain at the peak stress for the concrete specimens with 2% air content is 0.0033, while for those with 7% air content, it is 0.0035. Although the strain values at peak stress are similar, with only a slight difference of 0.0002, it is important to recognize that the overall mechanical behavior of the specimens is still influenced by the air content. Therefore, even though both materials may undergo similar amounts of plastic deformation at peak stress, the concrete with lower air content is still better suited for dynamic applications due to its overall superior compressive strength.

Table 2 The average mechanical properties of the tested specimens.

Air percent	Data	Strain rate (1/S)	Dynamic compressive strength (MPa)	Plastic strain at the peak stress	Density (kg/m^3)
2%	Avg.	14.24	70.58	0.0033	2368.20
	SD	0.74	2.07	0.0001	27.88
	CV, %	5.17	2.94	3.4975	1.18
7%	Avg.	17.86	59.00	0.0035	2288.00
	SD	1.11	0.38	0.0004	8.54
	CV, %	6.24	0.65	10.89	0.37

Conclusions

This study explores the impact of varying air content percentages on the compressive dynamic performance of conventional concrete. For this reason, cylindrical specimens with air content levels of 2 and 7% were prepared and subjected to dynamic strain rate testing. Based on the analyses, the following findings were drawn: The importance of carefully considering and optimizing air content percentages to enhance the dynamic performance of concrete. Also, reducing air content in concrete significantly enhances the dynamic compressive strength. However, air content variations showed no notable impact on the plastic strain at peak stress.

Conflict of interest statement No possible conflicts of interest were disclosed by the authors.

Data availability statement The corresponding author will send the datasets generated and/or analyzed in the current work upon reasonable request.

References

1. Thomas, C., J. Setién, J. Polanco, P. Alaejos, and M.S. De Juan, Durability of recycled aggregate concrete. Construction and building materials, 2013. 40: p. 1054-1065.
2. Shah, H.A., Q. Yuan, and S. Zuo, Air entrainment in fresh concrete and its effects on hardened concrete-a review. Construction and Building Materials, 2021. 274: p. 121835.
3. Bogas, J.A., J. De Brito, and D. Ramos, Freeze–thaw resistance of concrete produced with fine recycled concrete aggregates. Journal of Cleaner Production, 2016. 115: p. 294-306.
4. Amriou, A. and M. Bencheikh, New experimental method for evaluating the water permeability of concrete by a lateral flow procedure on a hollow cylindrical test piece. Construction and building materials, 2017. 151: p. 642-649.
5. Shi, Y., H. Yang, S. Zhou, A. Wang, and X. Lv, Effect of atmospheric pressure on performance of AEA and air entraining concrete. Advances in Materials Science and Engineering, 2018. 2018(1): p. 6528412.
6. Peng, G.-F., Q. Ma, H.-M. Hu, R. Gao, Q.-F. Yao, and Y.-F. Liu, The effects of air entrainment and pozzolans on frost resistance of 50–60 MPa grade concrete. Construction and Building Materials, 2007. 21(5): p. 1034-1039.
7. Tanesi, J. and R. Meininger, Freeze-thaw resistance of concrete with marginal air content. Transportation research record, 2007. 2020(1): p. 61-66.
8. Hang, M., L. Cui, J. Wu, and Z. Sun, Freezing-thawing damage characteristics and calculation models of aerated concrete. Journal of Building Engineering, 2020. 28: p. 101072.
9. Richardson, A., K. Coventry, and J. Bacon, Freeze/thaw durability of concrete with recycled demolition aggregate compared to virgin aggregate concrete. Journal of Cleaner Production, 2011. 19(2-3): p. 272-277.
10. Heard, W., B. Martin, X. Nie, T. Slawson, and P. Basu, Annular pulse shaping technique for large-diameter Kolsky bar experiments on concrete. Experimental Mechanics, 2014. 54: p. 1343-1354.
11. Shemirani, A.B., R. Naghdabadi, and M. Ashrafi, Experimental and numerical study on choosing proper pulse shapers for testing concrete specimens by split Hopkinson pressure bar apparatus. Construction and Building Materials, 2016. 125: p. 326-336.
12. Abbas, M.A., B.A. Williams, W.F. Heard, and X. Nie, Residual mechanical properties of high strength concrete BBR9 subjected to dynamic uniaxial compressive loading. Cement and Concrete Research, 2024. 180: p. 107488.
13. Chen, X., L. Ge, J. Zhou, and S. Wu, Experimental study on split Hopkinson pressure bar pulse-shaping techniques for concrete. Journal of Materials in Civil Engineering, 2016. 28(5): p. 04015196.
14. Ge, X., Y. Ge, Q. Li, X. Cai, W. Yang, and Y. Du, Effect of low air pressure on the durability of concrete. Construction and Building Materials, 2018. 187: p. 830-838.

Chapter 2
Energy Dissipation and Shock-Induced Phase Changes in Mock Energetic Crystals via In-Situ Raman Imaging

Mahavir Singh*, **Seongmin Yoon, and Vikas Tomar**

Abstract This study investigates shock deformation and energy dissipation in mock-energetic crystals using advanced full-field time-gated Raman spectroscopy. The experimental approach involved multiple timestamps to assess shock dissipation along the [1 0 0] direction, aligning with the direction of shock propagation. A total of 1200 experiments were conducted to integrate and construct Raman spectra over four timestamps, each of a duration of short time windows of less than 10 ns, allowing for detailed observation of shock-induced changes. Flyer velocities of 1.2 km/s were utilized to generate high-pressure conditions of up to 5 GPa and temperatures exceeding 1000 °C within the sample. The motion of the aluminum flyer was monitored, and its planarity was confirmed using high-speed optical imaging, which demonstrated the necessity of maintaining a uniform impact to prevent non-uniform deformation during shock propagation. Further energy density dissipation was computed to be $6.67 \pm 2.11 \times 10^7$ kJ/m^3 and was found evenly distributed in the shocked region. The shock loading caused formation of a metastable phase which disappeared leading to micromechanical crushing and spall type failures in the crystals. For an impact velocity of 1.2 km/s the shock velocity was found to be 6.25 km/s. This comprehensive analysis provides insights into the elastic-plastic transitions and phase changes occurring in solid materials and a method to determine shock response in materials especially energetics which are opaque to PDV/VISAR wavelengths.

Keywords Shock deformation · Time-resolved Raman Spectroscopy · Energy dissipation · Energetic Crystals · Metastable Phase

Introduction

Understanding material response under shock compression is crucial for aerospace, defense, and materials science applications. Extreme pressures and temperatures induce complex deformation mechanisms, including elastic-plastic transitions, energy dissipation, and phase transformations [1, 2]. Characterizing these processes is essential for designing materials that endure high-strain-rate conditions. Traditional shock compression studies use photon Doppler velocimetry (PDV), X-ray diffraction, and laser interferometry to measure stress-wave propagation, but these methods lack real-time molecular-level resolution. Ultrafast spectroscopic techniques, particularly time-resolved Raman spectroscopy, have advanced the study of shock-induced structural changes at atomic and molecular scales [3].

Organic crystalline materials, including energetic compounds and their mock equivalents, undergo significant chemical and structural changes under shock loading [4]. Internal energy redistribution, metastable phase formation, and molecular rearrangements impact mechanical integrity and failure modes. Raman spectroscopy enables in-situ observation of these transformations, providing insights into energy localization and dissipation [5]. Sucrose, a widely used model system for shock studies, shares structural similarities with polymer-bonded explosives and has well-characterized Raman-active modes [3, 5]. Shock compression induces distinct phase transitions in sucrose, marked by shifts in CH and CH2 vibrational modes, which help quantify local stress and temperature distribution [5, 6]. This study employs multi-point time-gated Raman spectroscopy to capture the spatial and temporal evolution of shock waves in sucrose crystals, integrating ultrafast optical diagnostics for characterization of energy dissipation and phase transformations.

Mahavir Singh · Seongmin Yoon · Vikas Tomar
School of Aeronautics and Astronautics, Purdue University, West Lafayette, IN 47907, USA
*Presenting Author: sing1042@purdue.edu
e-mail: yoon164@purdue.edu; tomar@purdue.edu

Paul Allison et al. (eds.), *Dynamic Behavior of Materials, Vol. 1*, Conference Proceedings
of the Society for Experimental Mechanics Series, https://doi.org/10.13052/9788743808275-2

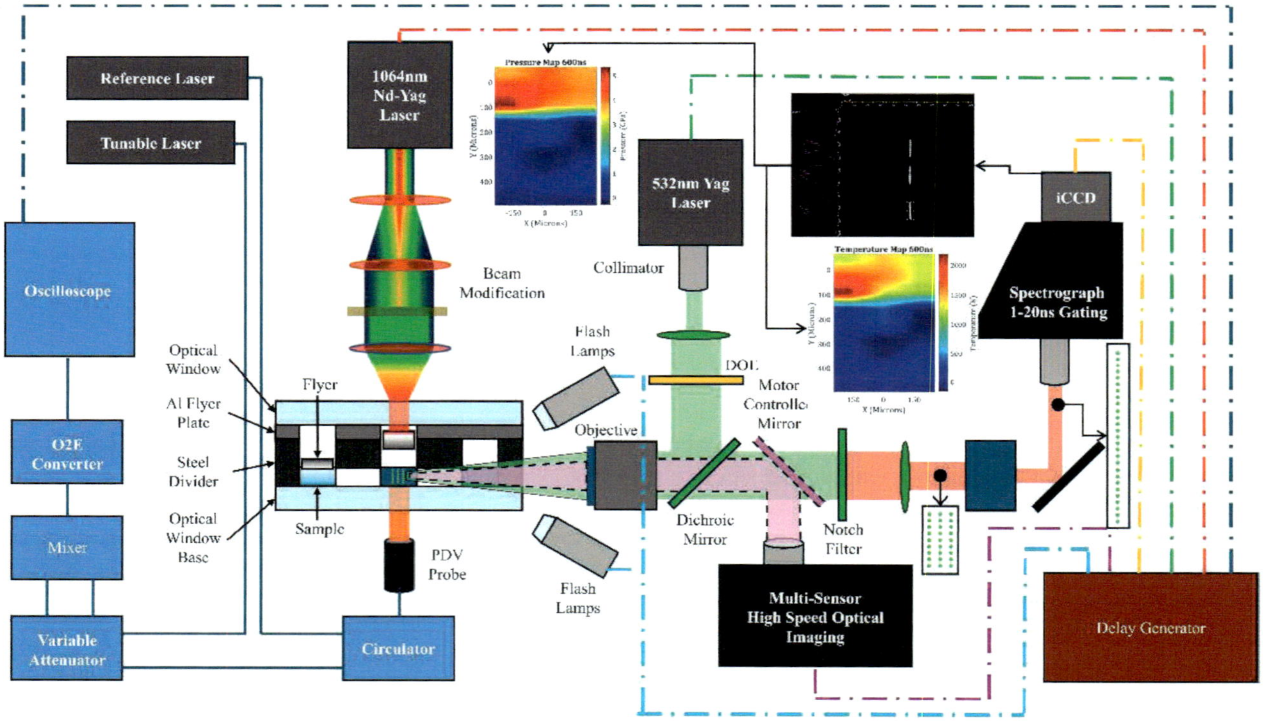

Fig. 1 Schematic of the experimental setup utilizing multi-point Raman spectroscopy to measure pressure and temperature distribution [6] in a shocked material, induced by a laser-driven flyer impact system. The peak shifts are subsequently used to calculate the energy dissipation density in the elastic shocked region.

Materials and Methods

Sucrose crystals were selected due to their well-characterized shock response and structural similarity to polymer-bonded explosives. The crystals were carefully polished and mounted to ensure uniform impact conditions. The experiments were conducted on samples with controlled thickness to achieve repeatable shock loading conditions. Shock waves were generated using a laser-driven flyer plate impact system, consisting of a 1064 nm pulsed laser with an 8 ns pulse width (figure 1). The specimens were mounted under a glass plate with an aluminum foil layer, from which aluminum flyers of 1.5 mm diameter were launched at velocities of 1.2 km/s. A precision alignment system ensured consistent impact positioning across multiple experiments.

Raman Spectroscopy and Optical Imaging: A custom-built multi-point, time-gated Raman spectroscopy system was employed to capture molecular-scale changes during shock loading. The system utilized a 532 nm pulsed laser with a 10 ns gate width. The laser beam was divided into 27 beams, which propagated in parallel and struck the front surface of the sample through a 532 nm dichroic mirror (figure 1). High-power switchable microscopic objectives enabled the acquisition of high-resolution Raman spectra at different spatial locations. After data collection, the 27 beams were optically modified to enter the spectrograph and an iCCD camera within a 10 ns gating window. The experiments were repeated 300 times to construct a complete Raman spectrum in the 2950–3050 cm^{-1} range, ensuring high-resolution data around the CH and CH2 intra-molecular vibration bands.

Results and Discussion

Raman spectra were recorded at four distinct timestamps corresponding to the undeformed state, elastic wave arrival, elastic-plastic deformation, and post-shockwave stages. The elastic peak shifts were analyzed to determine pressure and temperature variations, with the CH and CH2 peaks at 2990 and 3018 cm^{-1} used to characterize the shocked state of the material. These peaks were calibrated to quantify changes in pressure and temperature during shock compression.

Pressure and Temperature Distributions: Using Equation 1, the pressure and temperature distribution as shown in figure 1, within the shock wave were mapped by shifts in the CH2 symmetric and asymmetric stretching modes. The maximum recorded pressure was 5 GPa, with a corresponding temperature exceeding 1000 °C. The spatial distribution of temperature closely followed the pressure gradients, suggesting strong thermomechanical coupling during shock propagation. The constants C1-C4 are correlation constants for pressure and temperature against the peak shifts of the CH and CH2 peaks.

$$dP = \frac{C_4 \cdot d\omega_1 - C_3 \cdot d\omega_2}{C_1 \cdot C_4 - C_2 \cdot C_3} \quad dT = \frac{C_2 \cdot d\omega_1 - C_1 \cdot d\omega_2}{C_2 \cdot C_3 - C_1 \cdot C_4} \tag{1}$$

Shock-Induced Energy Dissipation and Phase Transitions: Utilizing the linear dependence of pressure and temperature given in Equation 1, and assuming the shock deformation is fast enough for the heat loss to be purely adiabatic in nature, the total energy dissipation density was derived using Equation 2.

$$E = \iint \frac{\partial^2 E}{\partial\omega_1 \partial\omega_2} d\omega_1 d\omega_2 = \frac{\rho_0}{C_3}\left(match\omega_1 - \omega_{1(0)}\right) \left[\frac{-bC_1 + \frac{aC_3}{K}}{C_2.C_3 - C_1.C_4}\left(\omega_2 - \omega_{2(0)}\right) + \frac{bC_1C_3}{K(C_1.C_4 - C_2.C_3)^2}\left(\omega_2^2 + 3\omega_{2(0)}^2 - 4\omega_2\omega_{2(0)}\right)\right] \tag{2}$$

The energy density was found to be 6.67 ± 2.11 × 107 kJ/m^3 within the elastic shocked region, with a uniform distribution. Additionally, Raman spectroscopy data revealed a sequential transformation of the material under shock loading. Initially, an elastic precursor wave propagated through the sucrose crystal, as indicated by small, reversible shifts in Raman peaks. As the shock wave intensified, a transition to an elastoplastic state was observed, characterized by peak broadening and the emergence of new vibrational modes. Subsequently, a metastable phase was observed, which disappeared after the shock wave passed. Under these deformations, the crystal underwent micro-mechanical crushing failure, converting into a pool of very fine powder. The fast and localized deformation caused the specimen to retain its original shape.

Conclusions

This work successfully demonstrates the capability of time-gated Raman spectroscopy in visualizing and quantifying shock-induced transformations in sucrose crystals. The experimental setup enabled high-fidelity measurements of pressure, temperature, and energy dissipation with unprecedented temporal and spatial resolution. The results contribute to a deeper understanding of elastic-plastic transitions and phase changes in mock energetic materials, providing a foundation for shock wave dynamics in energetic materials.

References

1. Coffey CS, Sharma J (1999) Plastic deformation, energy dissipation, and initiation of crystalline explosives. Phys Rev B 60:9365–9371
2. Harjono HL, Ravva S, Singh M, Huang H, Tomar V (2024) Data science assisted cohesive finite element modelling of impact behaviour of AP-HTPB crystal binder composite. Strain 60:e12465
3. Dhiman A, Lewis NS, Olokun A, Dlott DD, Tomar V (2022) Thermo-mechanical behavior measurement of polymer-bonded sugar under shock compression using in-situ time-resolved Raman spectroscopy. Sci Rep 12:1–9
4. Patterson JE, Dreger ZA, Gupta YM (2007) Shock wave-induced phase transition in RDX single crystals. Journal of Physical Chemistry B 111:10897–10904
5. Dhiman A, Dillard TA, Ehler A, Karmarkar S, Tomar V (2023) Visualizing Shock Induced Thermo-Mechanical Change at Bi-Crystal Interface Using Laser Array Based Nano-Second Raman Spectral Imaging. Journal of Dynamic Behavior of Materials 9:334–344
6. Singh M, Yoon S, Dhiman A, Dlott DD, Tomar V (2025) Measurement of Elastic-Plastic Shock Propagation and Phase Transition Using Multi-Point Time-Gated Raman Spectroscopy. Journal of Applied Physics

Chapter 3
Evaluation of Various Pulse Shaper Materials for Tension Split-Hopkinson Bar Experiments

Sydney C. Resnick and Nathan C. Spulak

Abstract The high velocity impact used to generate the stress-strain waves during a split-Hopkinson bar test can result in undesired oscillations in the incident wave, creating complicated loading and force imbalance on the specimen. Additionally, for some tests (such as on low-impedance or brittle materials), it is necessary to increase the rise time of the incident wave in order to achieve dynamic force equilibrium prior to significant deformation or specimen fracture. Pulse shapers consisting of a ductile material can be placed between the striker and incident bars in order to decrease these wave oscillations and increase the rise time. This allows dynamic force equilibrium to be more easily achieved, but often results in lower strain rate loading on the specimen. The effect of pulse shapers constructed from various different materials and thicknesses are investigated when used for tension split-Hopkinson bar tests. The tension test setup uses a hollow striker bar that impacts a flange on the end of the incident bar to generate the tension incident wave, and this setup necessitates the use of hollow ring-shaped pulse shapers attached to the flange. Pulse shapers of copper, nylon, and rubber materials with various thicknesses are used to evaluate the effectives of each pulse shaper in altering the oscillations, rise time, and pulse duration of the incident wave.

Keywords Split-Hopkinson bar · Pulse shaping · Tension · Elastic wave analysis

Introduction

Split-Hopkinson bar (SHB) tests are a widely utilized method for studying material behavior under high strain rate loading. Such information is necessary for applications such as automotive and aerospace vehicle crashworthiness, military and defense, and projectile or debris impacts. The SHB test, developed by Kolsky in the 1940s [1], involves a specimen placed between an incident and transmitter bar. A third striker bar is accelerated and impacts the incident bar, generating a stress-strain wave in the incident bar that then loads the specimen at high strain rates. However, due to the nature of the impact, the resulting stress waves can exhibit significant oscillations, leading to oscillatory loading on the specimen and potentially distorting the experimental data. Additionally, in some cases it is necessary to increase the rise time of the incident, such as for low impedance materials or brittle materials, in order to allow for dynamic force equilibrium to be achieved before significant deformation or fracture of the specimen occurs. To mitigate these oscillations and increase the rise time, pulse shapers may be employed to change the shape of the incident wave and improve the accuracy of the measured material properties [2]. The material properties, thickness, and shape of the pulse shaper will affect the behavior of the resulting incident wave. Copper, known for its high ductility and energy absorption capacity due to its ability to plastically deform, is a commonly used material for pulse shapers in SHB testing. Less commonly used materials include polymers and elastomers, such as nylon and rubber. This study investigates the effect of various different pulse shaper materials and thicknesses for tension SHB experiments utilizing a hollow striker bar impacting a flange on the end of a solid incident bar. Such a test setup requires utilizing ring-shaped pulse shapers as opposed to the more commonly used solid cylindrical pulse shapers.

Sydney C. Resnick · Nathan C. Spulak
Department of Mechanical and Aerospace Engineering University of Alabama in Huntsville, Huntsville AL 35899
email: sc1422@uah.edu; ncs0023@uah.edu

Paul Allison et al. (eds.), *Dynamic Behavior of Materials, Vol. 1*, Conference Proceedings of the Society for Experimental Mechanics Series, https://doi.org/10.13052/9788743808275-3

Experimental Methods

SHB tests are conducted in the tension configuration utilizing a hollow striker bar to generate the incident wave. The specimen is attached to both the incident and transmitter bar, and the hollow striker bar is accelerated using a pressure vessel to strike a flange on the end of the incident bar. Upon impact, a tensile stress wave is generated that travels down the incident bar to the specimen, at which point a portion of the wave is reflected back in the opposite direction along the incident bar as a compression wave, while the remainder of the wave is transmitted through the specimen and travels down the transmitter bar. Strain gages on the incident and transmitter bars measure the waves, which are subsequently converted to strain and stress values. The test setup is illustrated in Figure 1 (a).

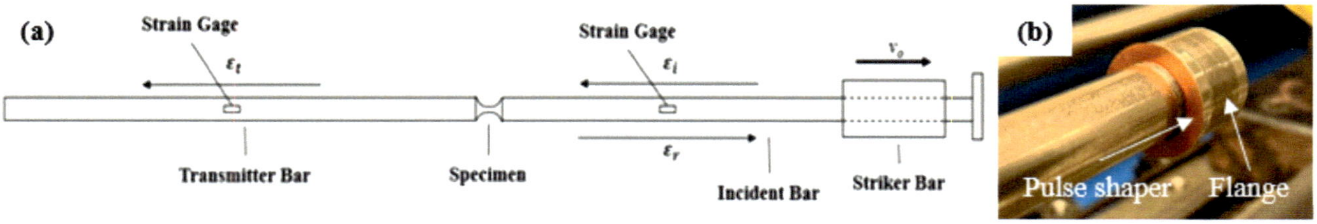

Fig. 1 (a) Experimental setup for a tension split-Hopkinson bar test, and (b) impact flange with ring-shaped pulse shaper

Pulse shapers are incorporated by removing the end cap of the incident bar and sliding a hollow shaper onto the end flange, as illustrated in Figure 1 (b), where it will interface with the striker bar upon impact. Copper pulse shapers with thicknesses of 0.005 inches, 0.045 inches, 0.09 inches, and 0.125 inches are used, as well as nylon pulse shapers with thicknesses of 0.062 inches and 0.25 inches and rubber pulse shapers of 0.13-inch thickness. The resulting effects on the incident wave rise time, duration, and oscillations (categorized by the initial percent overshoot) are analyzed for each pulse shaper material and thickness.

Experimental Results

The stress versus time data for the initial incident wave, generated using copper, nylon, and rubber pulse shapers of varying thicknesses are presented in Figure 2. The baseline data, recorded with no pulse shaper, serves as a reference for assessing the effects of the pulse shapers.

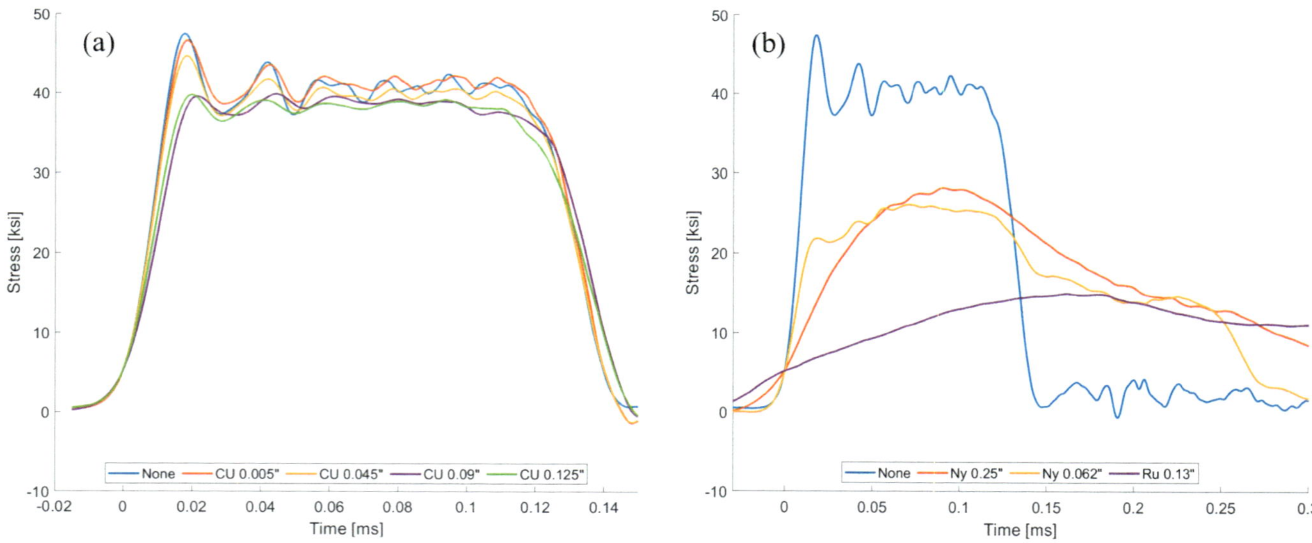

Fig. 2 Effect of various pulse shapers on incident wave: (a) copper pulse shapers, and (b) nylon and rubber pulse shapers

For the copper pulse shapers, the thickness had little impact on the wave duration. However, the thicker copper pulse shapers did reduce the amount of oscillations, as well as decrease the overall wave amplitude while still maintaining the overall square-wave shape. The nylon and rubber pulse shapers, by comparison, are able to increase the wave duration and significantly increase the rise time. Yet they also significantly decrease the wave amplitude and change the overall shape of the incident wave such that it no longer resembles a square wave.

These effects are numerically categorized by calculating the pulse duration, percent overshoot, and rise time of the wave for each pulse shaper type. The pulse duration is calculated by taking the peak measurement, halving it, and taking the duration from the point it initially reaches 50% of the peak to the point it falls back to 50%. The baseline duration when no pulse shaper is used is 122.10 μs and the most deviation from this baseline is seen when using the 0.09 inch shaper. The pulse shaper of 0.005 inches had a duration of 122.93 μs, a 0.686% increase. The 0.045-inch-thick pulse shaper had a duration of 122.89 μs, or an increase of 0.645%. The 0.09-inch-thick pulse shaper had a duration of 125.89 μs, or 3.1% increase. Finally, the 0.125-inch-thick pulse shaper had a duration of 125.66 μs, increasing by 2.92%.

The effect of the copper pulse shapers on the rise time is shown in Figure 3, which shows only the initial rise of the incident wave for each copper pulse shaper. The rise time of the wave is calculated as the time it takes for the wave to go from 10% to 90% of the peak value. The baseline with no pulse shaper has a rise time of 14.68 μs. The 0.005-inch-thick shaper has a rise time of 15.49 μs, increasing by 5.5%. The 0.045-inch-thick shaper produces a rise time of 15.19 μs, increasing by 3.5%. The 0.09-inch-thick shaper has a rise time of 17.7 μs, demonstrating a more significant increase of 20.5%. Lastly, the 0.125-inch-thick shaper had a rise time of 16.38 μs, an increase of 11.6%.

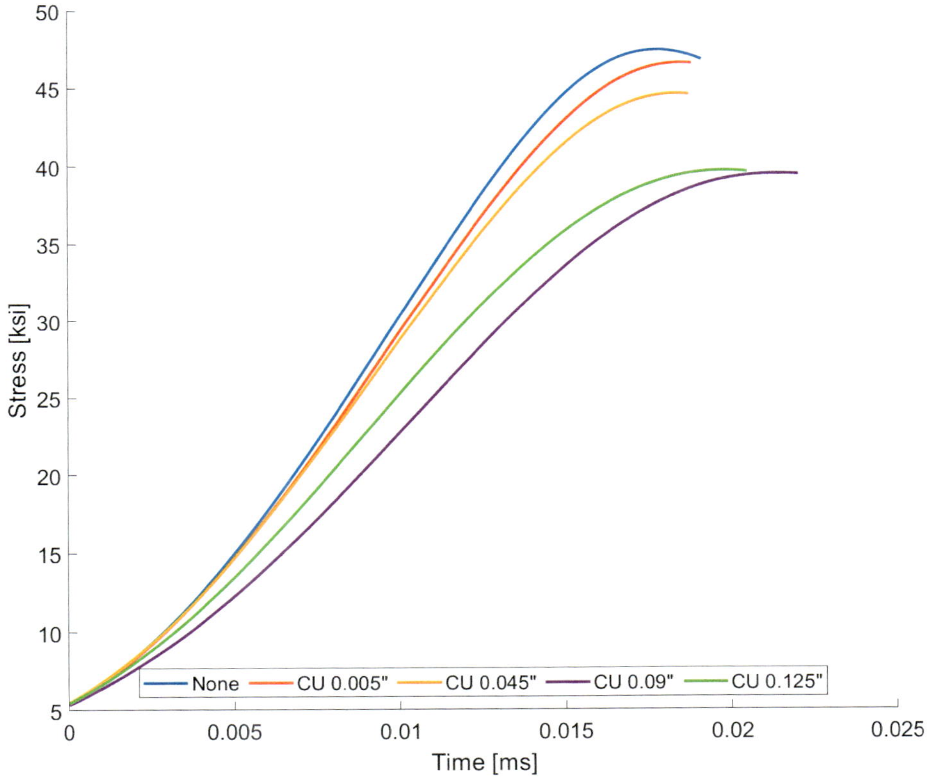

Fig. 3 Initial rise time of incident wave for various copper pulse shaper thicknesses

The percent overshoot of the wave is calculated by first determining the average value of the wave after the initial peak has subsided, in order to obtain a value of the steady-state wave amplitude. The percent overshoot is then determined as the percentage difference between the initial peak amplitude and the steady-state amplitude. With no pulse shaper, the wave had a 14.9% overshoot. The 0.005-inch-thick shaper saw an overshoot of 12%, the 0.045-inch-thick shaper had overshoot of 11.4%, the 0.09-inch-thick shaper had a significantly lower overshoot of 3.28%, and the 0.125-inch-thick shaper had an overshoot of 3.91%.

A linear trendline is fit to percent overshoot verses pulse shaper thickness for copper, as seen in Figure 4. This can be utilized for future testing, to select the appropriate pulse shaper thickness that results in the desired reduction in percent overshoot and wave oscillations.

In addition to copper, nylon and rubber pulse shapers are used, with notable differences observed in their effects on the incident wave. These materials result in more pronounced changes in wave characteristics, including a substantial reduction

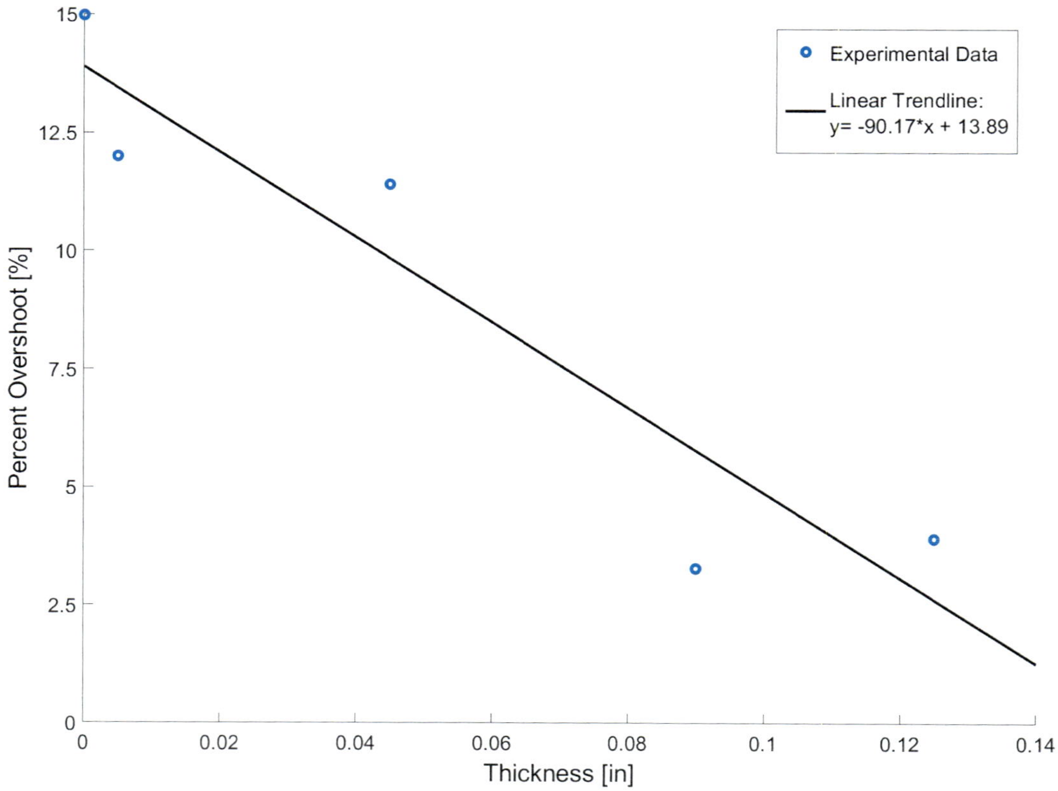

Fig. 4 Experimental data and linear trendline of percent overshoot as a function of pulse shaper thickness

in peak amplitude and pulse duration. These pulse shapers are a potential option for tests where an increased pulse duration or significantly increased rise time are required, such as tests on low impedance or brittle materials.

Conclusion

This study highlights the effect of various pulse shaper options on the incident waves generated by a hollow striker bar for tension split-Hopkinson bar tests. Copper pulse shapers, particularly those thicker than 0.05 inch, effectively reduced oscillations in the incident wave, resulting in lower peak values and overshoot. However, they had a much smaller effect on the rise time and minimal effect on wave duration. In contrast, nylon and rubber pulse shapers led to more substantial changes in wave shape, duration, and overshoot, making them potentially better suited for tests involving brittle or low impedance materials which require increased rise times. These findings underscore the importance of selecting the appropriate pulse shaper material and thickness for specific experimental needs, to achieve more accurate and reliable high-strain-rate testing.

Acknowledgments This research was supported by the NASA Marshall Space Flight Center (MSFC) Cooperative Agreement Notice (CAN) 80NSSC24M0075.

References

1. Kolsky, H. (1949). An investigation of the mechanical properties of materials at very high rates of loading. *Proceedings of the physical society. Section B*, *62*(11), 676.
2. Gray III, G. T. (2000). Classic split-Hopkinson pressure bar testing. In *Mechanical testing and evaluation* (pp. 462-476). ASM International.
3. Seidt, J. D. (2010). Plastic deformation and ductile fracture of 2024-T351 aluminum under various loading conditions (Doctoral dissertation, The Ohio State University).

Chapter 4
Surface Deformation Measurements of a Composite Cylinder Composed of Particulate Idoxuridine Crystals Embedded in Estane Under Compression

Pooyan B. Javadzadeh, Ehsan Mehrdad, Yao Ren, Nathan Peterson, Amy Clarke, Hongbing Lu

Abstract 5'iodo'2 idoxuridine has served as a reliable replacement for plasticly bonded explosives (PBX) in university environments to characterize mechanical behavior under quasi-static and dynamic testing conditions. Particulate idoxuridine crystals with particle sizes in the range of 75-150 μm are embedded in Estane polymer matrix to form a particulate composite cylinder specimen of 5 mm diameter. A metallic ink marker is applied to the surface of the composite cylinder, and volumetric images of the specimen are obtained using micro-computed tomography (μCT) at approximate 4 $\mu m/pixel$ resolution to determine the internal microstructure, including any voids. The metallic ink markers are distinguishable in each μCT volumetric image, and in the surface images of the same specimen, allowing for the correlation of internal microstructure to the surface speckles. The μCT pre-scanned specimen is compressed under quasi-static loading condition, where a stereo-microscope with a resolution of 4.0 μm/pixel is used to acquire the surface stereo images of the specimen undergoing deformation. The surface stereo images are analyzed using stereo (or 3D) digital image correlation to determine the surface axial, circumferential, and shear strains. These strain results are correlated to the internal microstructure for validation of meso-scale simulations of the specimen under compression. An 80-feet-long split Hopkinson pressure bar is used to apply compression while a high speed camera is used to acquire surface images at ambient temperature. Digital image correlation is used to determine the surface strains of the specimen. The markers again allow for correlation of the surface deformations to the microstructure. These experimental results provide detailed internal microstructure of the particulate composite specimen, which are correlated to the measured surface strains under both quasi-static and dynamic loading conditions. The results can be used for verification and validation of simulations of the mechanical response of particulate materials by direct numerical simulation (DNS), and to assist in the development of a computational multiscale micromorphic continuum mechanics framework.

Keywords Granular mechanics · Particulate composites · Micro-computed tomography · Digital image correlation · IDOX/Estane

Introduction

Numerous surrogate materials have been used for energetic particles in the plastic-bonded explosives (PBX. These include pentaerythritol, acetaminophen, and sucrose [1], [2], [3]. Each of such inert materials has its advantages and shortcomings. Yeager et al. [4] investigated multiple candidates with properties similar to HMX, these included IDOX, MDTTO, MET-I, N-BPFPO, PFBA, and PFBBA, they found that IDOX, N-BPFPO, and PFBA showed similar processability characteristics and thermal behavior. Later, Yeager et al. [5] investigated the thermal stability, density, crystal structure, and processing characteristics of IDOX, N-BPFPO, and PFBA, and concluded that the IDOX-based mock is less water-soluble than current

Pooyan B. Javadzadeh · Ehsan Mehrdad · Yao Ren · Hongbing Lu
Department of Mechanical Engineering, The University of Texas at Dallas, Richardson, TX 75080
e-mail: pooyan.javadzadeh@utdallas.edu; Ehsan.Mehrdad@utdallas.edu; Yao.ren@utdallas.edu; amyclarke@mines.edu

Nathan Peterson
Department of Metallurgical and Materials Engineering, Colorado School of Mines, Golden, CO 80401
e-mail: nepeterson@mines.edu

Paul Allison et al. (eds.), *Dynamic Behavior of Materials, Vol. 1*, Conference Proceedings of the Society for Experimental Mechanics Series, https://doi.org/10.13052/9788743808275-4

surrogate materials. Burch et al. [6] performed nanoindentation on IDOX crystals to investigate the mechanical properties of the powder and found that IDOX and HMX crystals have similar properties (e.g., a reduced modulus of 23.3 GPa for IDOX versus 25.2 GPa for HMX, and a hardness of 1.0 GPa for both), concluding that IDOX exhibits similar elastic and plastic mechanical behavior to HMX. Energetic materials typically contain approximately 95% explosive and 5% matrix or binder. The particle size distribution is often 50 wt.% fine ($<75\ \mu m$) and 50 wt.% coarse ($>150\ \mu$m). Historically, Estane 5703 polymer has served as the matrix in HMX and PBX composites. The mechanical properties of composite IDOX embedded in Estane have been investigated under quasi-static and high-strain rate conditions (e.g., using a split Hopkinson pressure bar) in [7], [8], [9]. In this study, we investigate the surface deformation of IDOX/Estane composite specimens—each 5 mm in both height and diameter (1:1 aspect ratio)—using a novel technique. First, the specimen is marked with conductive ink (with verification that the ink does not alter its mechanical properties, as confirmed by comparison with unmarked specimens). The marked specimen is then scanned, and stereo DIC is performed under quasi-static loading to study full-field surface transformation and deformation. The same method is applied under SHPB conditions using 2D DIC. This approach aids in aligning the DIC data with the pre-scanned specimen, thereby facilitating post-test segmentation for modeling and experimental verification.

Materials

5'iodo'2 idoxuridine (IDOX) particles of (75-150 μm) were mixed with Estane, the IDOX crystals obtained from Los Alamos National Laboratory (LANL), were recrystallized. The mixture was stirred at 50 °C until fully dissolved and then transferred to a containment vessel for crystallization, The crystallized IDOX was filtered using ultrafine mesh coffee filter paper, and any remaining IDOX was washed off the filter using tetrahydrofuran (THF).

After drying, percussive grinding was used to break the larger crystals into smaller particles. These particles were mixed with Estane resin at a ratio of particulate to matrix of 19:1 by weight. The mixture was placed into a preheated die (90 °C) and compressed under 155 MPa for one minute.

Experiments

Quasi-static Compression and 3D DIC

Quasi-static compression tests were conducted in compliance with ASTM D695 standard using an Instron 5969 universal material testing machine equipped with a 500 N load cell. The experiment was performed under ambient temperature

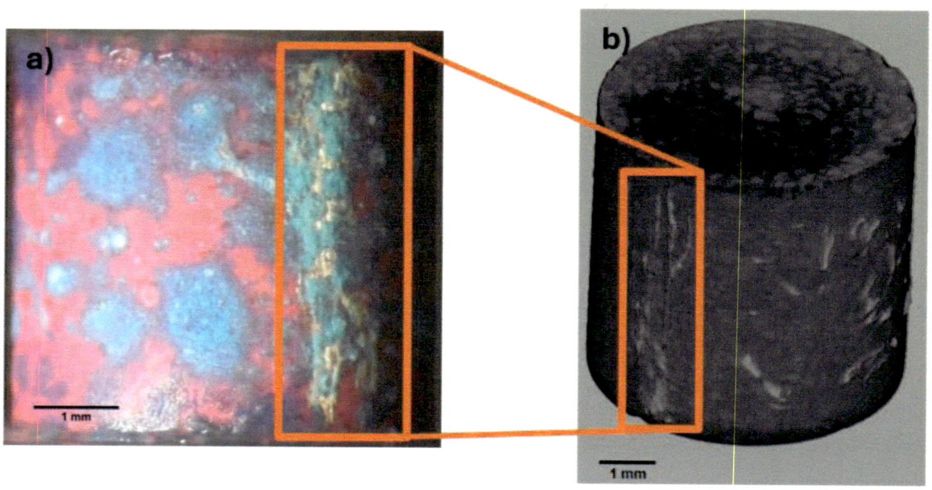

Fig. 1 (a) Representative of stereo DIC set up with a Zeiss microscope and Nikon 7200 cameras, (b) speckled 90 °C press IDOX/Estane with visible inks under X-ray computed tomography

conditions of 23 °C and 44% relative humidity. A Zeiss long working distance stereo microscope (Model 310187) was fitted with a pair of Nikon D7200 digital cameras (at 6000×4000 pixel resolution) each attached with a 105 mm lens, was used to acquire images of the specimen as the specimen deforms. First the specimen was painted with non-reflective water base paint to avoid glares and reflection of LED lights for better correlation results, Petroleum jelly (Vaseline) was applied to both ends of the cylindrical specimen to reduce surface friction as outlined here [10] then the specimen was carefully placed at the center of 2" platen compression and compressed at the rate of 0.5 mm/min. The specimen was placed in a pre-determined orientation and in order to make post processing data easier after 3D DIC analysis to ensure displacement fields match the acquired data from the machine and to match the grains in μCT. Fig. 1 (a) shows a 5 mm in height and diameter with applied speckles and conductive ink and Fig. 1 (b) shows corresponding marked strip visible under X-ray computed tomography (CT) with rendering 4 $\mu m/pixel$.

High strain rate compression 2D DIC

An 80-ft long SHPB was used to perform high strain rate compression on a pre-scanned 5 mm in height and diameter IDOX/Estane composite. A Photron high speed camera capable of achieving 120,000 fps was used with a 105 mm lens. Pulse shaping technique was adapted to achieve dynamic force equilibrium and constat strain rate during high rate of loading. For a striker of 457 mm, the duration of the experiment is 481 μs. In a typical SHPB experience transmitted wave is used to calculate stress, and reflected wave is used to calculate strain rate, as the wave travels in bars sufficient amount of time is required for the specimen to achieve equilibrium state; pulse shaping has proven to be reliable method in achieving equilibrium. In order to acquire material properties in dynamic testing, nearly uniform deformation is required. In a SHPB stress (σ_s) and strain rate ($\dot{\varepsilon}_s$) are calculated using equations (1) and (2), respectively.

$$\sigma_s = \frac{EA}{A_s}\varepsilon_t \tag{1}$$

$$\dot{\varepsilon}_s = \frac{-2c}{l_0}\varepsilon_r \tag{2}$$

where, E, A, and c are Modulus of elasticity, cross-sectional area, and elastic bar wave velocity, respectively and l_0 *and* A_s are the original length and cross-sectional area of the specimen respectively.

Figure 2 shows pre-scanned marked specimen with conductive inks which are visible to X-ray without changing the geometry of the specimen. During high strain rate experiment, the randomly applied markers were used as speckles on the surface of the specimen to measure surface deformation during SHPB experiment. Overall, two specimens were tested under the same condition at 550 s^{-1} strain rate, with overlapping results.

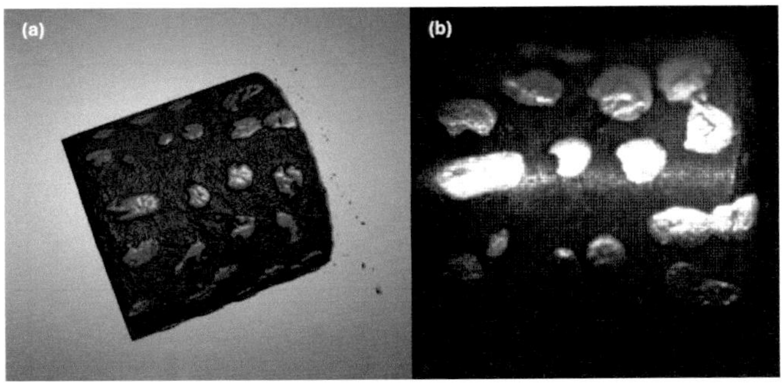

Fig. 2 Corresponding conductive ink markers (a) micro-CT of 5 mm IDOX/Estane specimen (b) same specimen placed in SHPB before impact

Results and Discussion

In this study we conducted quasi-static and high strain rate experiments on 5 mm in height and diameter IDOX/Estane specimens manufactured at 90 °C and used 3D and 2D DIC to acquire surface deformation during quasi-static and high strain rate testing, respectively.

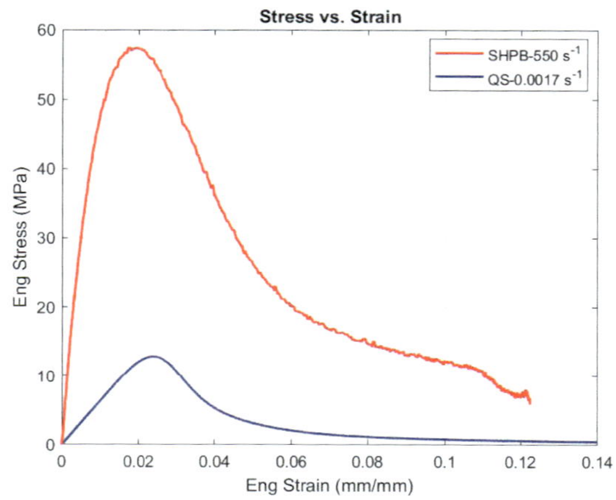

Fig. 3 Representative of stress-strain relationship of 5 mm IDOX/Estane during quasi-static at 0.0017 s-1 strain rate (blue) and stress-strain relationship under uniaxial SHPB experiment at 550 s-1 strain rate (red).

We tested 5 specimens of 5 mm in height and diameter of composite IDOX/Estane to ensure repeatability of the data. Fig. 3 shows stress-strain relationships of both quasi-static and high strain rate experiments of 5 mm specimen at room temperature. Since the margin of error for all the tests during quasi-static and medium strain rate was withing 5%, only one plot of each experiment is presented. As shown in Fig. 3 both specimens reach max stress during low and medium strain rate at almost the same strain, following a linear region and after which reflects the plastic region. It can be seen that the stress-strain curve indicates the behavior of soft materials, there is pre-peak region corresponding to elastic section following immediately by plastic region.

Fig. 4 illustrates 2D DIC acquired during SHPB, since crystal structure of IDOX is triclinic with space group of P1, and since triclinic lattice structure lacks symmetry elements, this makes the composite IDOX/Estane highly anisotropic. The

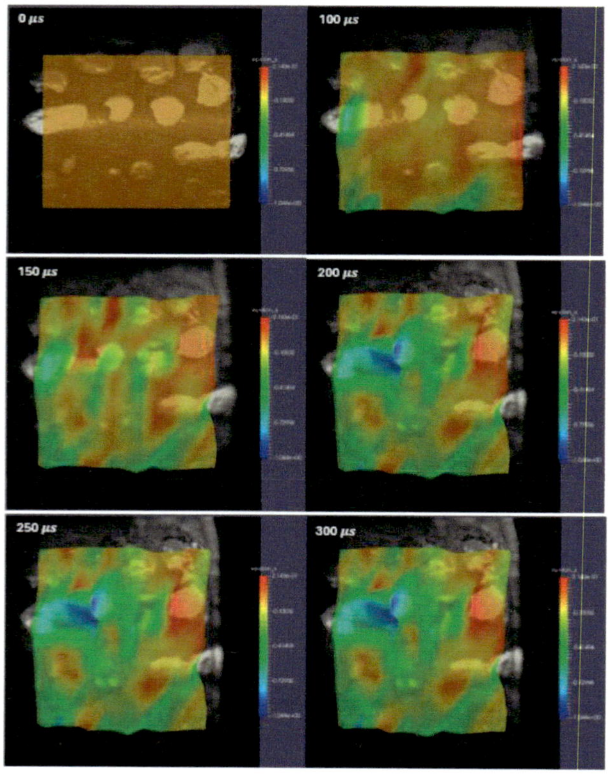

Fig. 4 Illustration of 2D DIC during high strain rate testing on 5 mm specimen

strain patterns are not radially symmetric, concluding the material does not respond identically in all direction, this suggests the composite specimen has different mechanical properties along different orientations. In general, composite materials exhibit anisotropic mechanical properties due to crystal structure. As the specimen deformed uniformly, the distribution of strain in the SHPB test suggests different stiffness and compliance in different regions. A total of two specimens were tested under the same conditions and approximately 550 s^{-1} strain rate, both materials showed similar behavior. This method helps in aligning data obtained during DIC to match the pre-scanned sample, to further assist simulation of mock energetic materials using high performance computers (HPC). The results can be utilized to verify and validate simulations of the mechanical behavior of particulate materials through direct numerical simulation (DNS) and to support the development of a computational multiscale micromorphic continuum mechanics framework.

Acknowledgments We acknowledge the support of the Department of Energy, National Nuclear Security Administration, Predictive Science Academic Alliance Program (PSAAP) under Award Number DE-NA0003962. We also appreciate support from the NSF under Award Numbers 2219347 and 1726435. H. Lu acknowledges the Louis A. Beecherl Jr. Chair for additional support. We are grateful to Dr. Alexandra Burch at Los Alamos National Laboratory for providing the crystal IDOX used in this work.

References

1. D. Eilerman and R. Rudman, “Refinement of pentaerythritol,” Acta Crystallogr B Struct Sci, vol. 35, no. 10, pp. 2458–2460, Oct. 1979, doi: 10.1107/S056774087900964X.
2. J. M. Welton and G. J. McCarthy, “X-Ray Powder Data for Acetaminophen (C8 H9 NO2),” Powder Diffr., vol. 3, no. 2, pp. 102–103, Jun. 1988, doi: 10.1017/S0885715600013348.
3. G. M. Brown and H. A. Levy, “Sucrose: Precise Determination of Crystal and Molecular Structure by Neutron Diffraction,” Science, vol. 141, no. 3584, pp. 921–923, Sep. 1963, doi: 10.1126/science.141.3584.921.
4. J. D. Yeager, A. L. Higginbotham Duque, M. Shorty, P. R. Bowden, and J. A. Stull, “Development of inert density mock materials for HMX,” Journal of Energetic Materials, vol. 36, no. 3, pp. 253–265, Jul. 2018, doi: 10.1080/07370652.2017.1375049.
5. J. D. Yeager et al., “Development of a New Density and Mechanical Mock for HMX”.
6. A. Burch, J. Yeager, and D. Bahr, “Nanoindentation of HMX and Idoxuridine to Determine Mechanical Similarity,” Crystals, vol. 7, no. 11, p. 335, Nov. 2017, doi: 10.3390/cryst7110335.
7. A. C. Burch, V. K. Euser, B. M. Morrow, C. M. Cady, and J. D. Yeager, “Compressive strength of the plastic-bonded explosive surrogate idoxuridine under quasistatic and dynamic compression,” Journal of Composite Materials, vol. 56, no. 26, pp. 3949–3956, Nov. 2022, doi: 10.1177/00219983221125904.
8. C. Liu, C. S. Woznick, S. J. Scott, A. L. Duque, M. J. Herman, and J. D. Yeager, “Brazilian disk compression testing of plastic-bonded idoxuridine mock explosive,” Journal of Energetic Materials, vol. 41, no. 3, pp. 319–350, Jul. 2023, doi: 10.1080/07370652.2021.1968073.
9. M. J. Herman, C. S. Woznick, S. J. Scott, J. T. Tisdale, J. D. Yeager, and A. L. Duque, “Composite binder, processing, and particle size effects on mechanical properties of non-hazardous high explosive surrogates,” Powder Technology, vol. 391, pp. 442–449, Oct. 2021, doi: 10.1016/j.powtec.2021.06.009.
10. W. Chen and B. Song, Split Hopkinson (Kolsky) Bar: Design, Testing and Applications. in Mechanical Engineering Series. Boston, MA: Springer US, 2011. doi: 10.1007/978-1-4419-7982-7.

Chapter 5
Strain Rate and Temperature Effect on the Dynamic Tensile Response of Carbon/Epoxy Composite

Palak Bhagoria, Muddu Rahul Bharadwaj, Rohit Kumar, and Vikrant Tiwari

Abstract In high-performance applications such as aerospace and automotive sectors, the dynamic behaviour of materials are crucial as these are often subjected to both high strain rate and varying temperature. This paper discusses the dynamic tensile performance of plain weave carbon/epoxy composite at high and low temperatures. Carbon/epoxy composite laminates were fabricated using the resin transfer moulding technique. The high strain rate experiments were conducted on tensile Split Hopkinson Pressure bar (TSHPB) in the strain rate range of 900 - 2700 s^{-1}. A furnace and an in-house built environment chamber along with liquid nitrogen was utilised for conducting high and low temperature studies, respectively. The temperature range varied from -50 °C to 60 °C. The mechanical properties including strength and strain at peak stress is evaluated and their variation with varying strain rate and temperature is examined. The strength was observed to enhance with increasing strain rates, however it degraded with rising temperatures. Failure mechanisms such as fiber breaking and matrix cracking have been observed, indicating the dynamic response of plain weave carbon/epoxy composite.

Keywords High strain rate · Temperature · Hopkinson pressure bar

Introduction

Carbon-based polymeric composites have gained popularity in recent decades due to their exceptional properties such as lightweight characteristics, high strength, excellent chemical resistance, and superior thermal stability. The growing demand for high-performance materials necessitates a deeper understanding of the properties, behaviour, and applications of carbon-based polymeric composites.

Furthermore, numerous real-world applications are characterized by high strain rates as well as elevated or varying temperatures. For example, elements in the aerospace or automobile fields encounter coupled thermal and mechanical stresses, where both the rate of strain and temperature significantly influence the material's behaviour. Thus, several researchers have made efforts to study the mechanical behaviour of materials under the influence of varying temperatures and strain rate. Most of such components are made up of different composite materials and hence, it becomes a necessity to evaluate their performance under such extreme conditions.

Several researchers have analysed the effect of high strain rate on different composites utilising split Hopkinson pressure bar(SHPB) along with techniques such as digital image correlation [1–3] , however only few focussed on the coupled effect of strain rate and temperature. Hosur et al. [4] examined plain and satin woven fabric graphite/epoxy composite at high strain rate compression at ambient and high temperatures using SHPB. They found peak stress and modulus to increase with increasing strain rates at all temperatures. However, temperature reduced peak stress and dynamic modulus. Temperature, strain rate, and fabric architecture also affected failure modes. Zhang et al. [5] investigated the dynamic mechanical properties and failure characteristics of AFRP (Kevlar 29/49 fabric and epoxy composite) at various medium strain rates (25,50,100, and 200 s^{-1}) and temperatures (− 25, 0, 25, 50, and 100 °C) and potential coupling effects of strain rate and temperature. The Young's modulus and tensile strength of both kinds of AFRP increased from 25 to 50 s^{-1} but declined at 200 s^{-1}. For AFRP K29, a similar pattern between temperature and Young's modulus was noted. Nonetheless, AFRP K49's Young's modulus remained unaffected by temperature changes. Another researcher [6] focussed on the evaluation

Palak Bhagoria · Muddu Rahul Bharadwaj · Rohit Kumar · Vikrant Tiwari
Department of Applied Mechanics Indian Institute of Technology Delhi, New Delhi, 110016, India
email: palakbhagoria@gmail.com; rahulbharadwaj95@gmail.com; rohitnituk001@gmail.com; Vikrant.Tiwari@am.iitd.ac.in

Paul Allison et al. (eds.), *Dynamic Behavior of Materials, Vol. 1*, Conference Proceedings of the Society for Experimental Mechanics Series, https://doi.org/10.13052/9788743808275-5

of the high strain rate compressive behaviour of cross-ply carbon/epoxy composite in the temperature range of 20 – 80 °C and observed marginal increase in ultimate compressive strength along with significant decrease in ultimate strain in the strain rate range of 10^{-3} to 10^{3} s^{-1} in all the three principal material directions. Also, the ultimate compressive strength decreased with increasing temperature. Similarly, other study [7] utilised SHPB experiments to analyse the dynamic mechanical performance of T700/BA9916 carbon/epoxy composite at high temperatures (25 – 200 °C). Elastic modulus and compression failure strength both declined with rising temperature, although maximum failure strain increased. Elastic modulus and compression failure strength both increased with increasing strain rates, while maximum failure strain decreased. Reis et al. [8] performed tensile experiments on glass fiber reinforced polymeric composite at various strain rates and temperatures and found that strain rate had a significant influence on ultimate tensile strength, but temperature had an effect only on modulus. Zhang et al. [9] examined the combined influence of temperature and strain rate on the tensile properties of basalt-fiber-reinforced polymers. The tensile strength remained unchanged at temperatures below 50 °C but dropped at 100 °C. Joshi et al. [10] conducted experimental study to characterize different composites under tensile loading for strain rates of 10^{-3}, 10^{-2}, and 10^{-1} s^{-1} and at temperatures 250 °C and 450 °C. The tensile behaviour of all the specimens indicated a significant influence of strain rate on ultimate tensile strength, whereas the temperature primarily affected the modulus.

To the best of author's knowledge, the literature reflects a lack of comprehensive study performed on plain weave carbon/epoxy composite with regard to the influence of high strain rate and varying temperatures, and thus an effort has been made to understand the dynamic response of this material under such extreme conditions. The study is performed for high strain rate tensile loading condition combined with both low and high temperature utilizing tensile split Hopkinson pressure bar (TSHPB) along with furnace and an in-house built environment chamber for introducing temperature effects.

Composite Fabrication and specimen Preparation

The composite laminates of thickness 2 mm are fabricated using 12 layers of plain weave carbon fabric and epoxy resin system (EPOFINE®-1564 and FINEHARD-3486-2) utilizing Vacuum Assisted Resin Transfer Moulding machine. The laminate was cured at room temperature for 24 hrs followed by further curing at 70 °C for 16 hrs. Dog bone specimens of 2 mm thickness, and gauge length and gauge width as 4 mm each, are obtained from these laminates using water jet cutting.

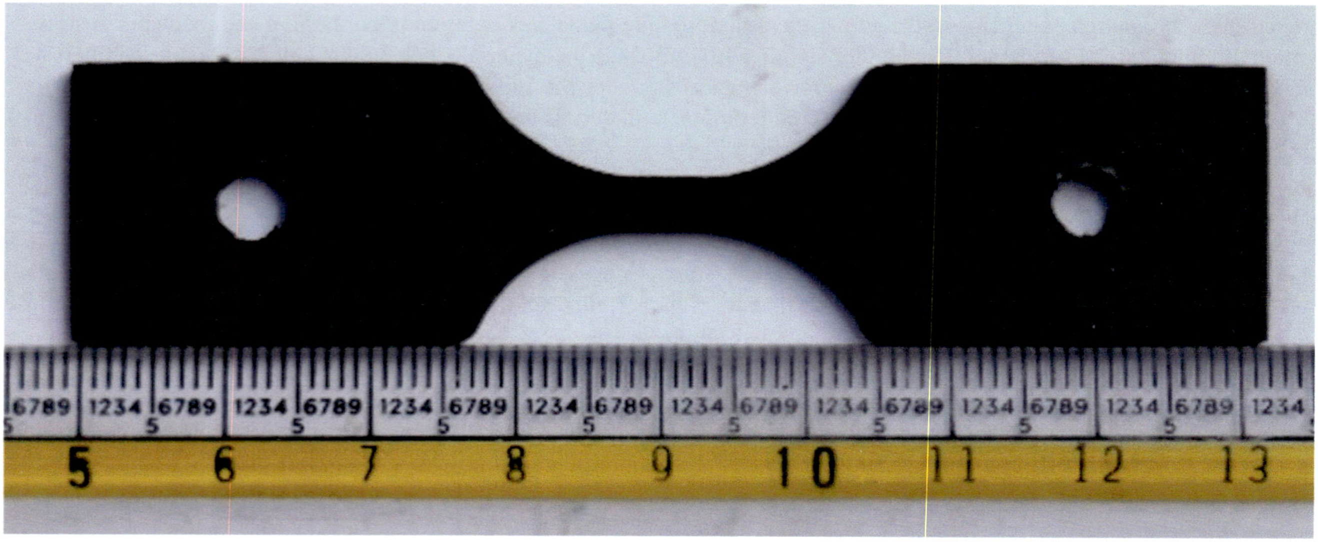

Fig. 1 Dog-bone shaped specimen for high strain rate tensile experiments

Experimental Setup

The high strain rate tensile experiments are performed on tensile split Hopkinson pressure bar (TSHPB) setup available in Dynamic Material Response and Characterization Laboratory in Applied Mechanics department of IIT Delhi. An elaborative description of the complete setup is presented in [11, 12] .This setup works on the principle of one-dimensional wave propagation theory, which thereby is used for the calculation of variation of stress, strain rate and strain with respect to time using equations (1), (2) and (3), respectively [13]:

$$\sigma_s\,(t) = \frac{A_b\,E_b}{A}\varepsilon_t(t) \tag{1}$$

$$\dot{\varepsilon} = \frac{2C_b}{L}\varepsilon_r \tag{2}$$

$$\varepsilon = \frac{2C_b}{L}\int_0^t \left[\varepsilon_r\,(t)\right]\,dt \tag{3}$$

where A_b and A are the cross-sectional areas of bar and specimen, respectively. C_b and E_b are the elastic wave speed and elastic modulus of bar. ε_r and ε_t are the strains of the reflected and transmitted pulse, respectively.

In order to perform experiments with flat specimens, the setup was modified by utilizing adapters made up of same material as the bar. The details of these adapters are well described in [14]. Aluminium tabs are pasted on the gripping length of the specimen to prevent any slippage and to protect specimen from any damage when fitted in the adapters.

The current study also included evaluation of the influence of varying temperatures, for which number of iterations for time versus temperature cycle were performed. These iterations, as shown in fig. 2, were utilised to perform experiments at desired temperature at a particular time. For attaining high temperatures, the specimen was heated in furnace at 100 °C for 1 hr and then subjected to high strain rate at different time instants. The low temperature experiments were conducted utilizing an in-house built environment chamber along with liquid nitrogen.

The study initiated by performing time versus temperature studies wherein the specimen is mounted in the adapter within the chamber and using K-type thermocouple, the variation of temperature with respect to time is noted. Number of iterations, as shown in fig. 2, were performed to evaluate the time at which the desired temperature is attained.

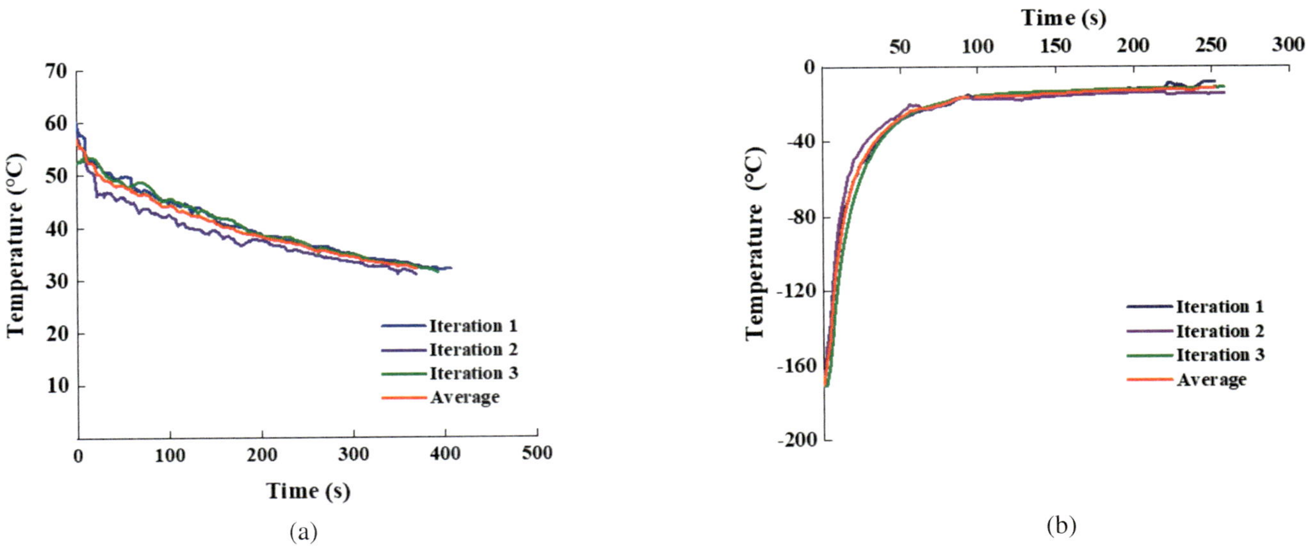

Fig. 2 Time versus Temperature curves for (a) High temperature, (b) Low temperature studies

Results and Discussion

High strain rate tensile experiments for plain weave carbon/epoxy composite were performed utilizing tensile split Hopkinson pressure bar in the strain rate range of 900-2700 s^{-1} at temperatures of − 45 °C, − 25 °C, 25 °C (room temperature),

35 °C and 45 °C. Fig. 3 represents the true stress versus true strain plots for plain weave carbon/epoxy composite corresponding to different temperatures for various strain rates. Table 1 enlists the strength and strains at peak stress obtained at these strain rates and temperatures. It can be noted that at any temperature, the strength and strain at peak stress increase with increasing strain rate, indicating strain rate sensitivity of plain weave carbon/epoxy composite (Fig. 4). The table also shows

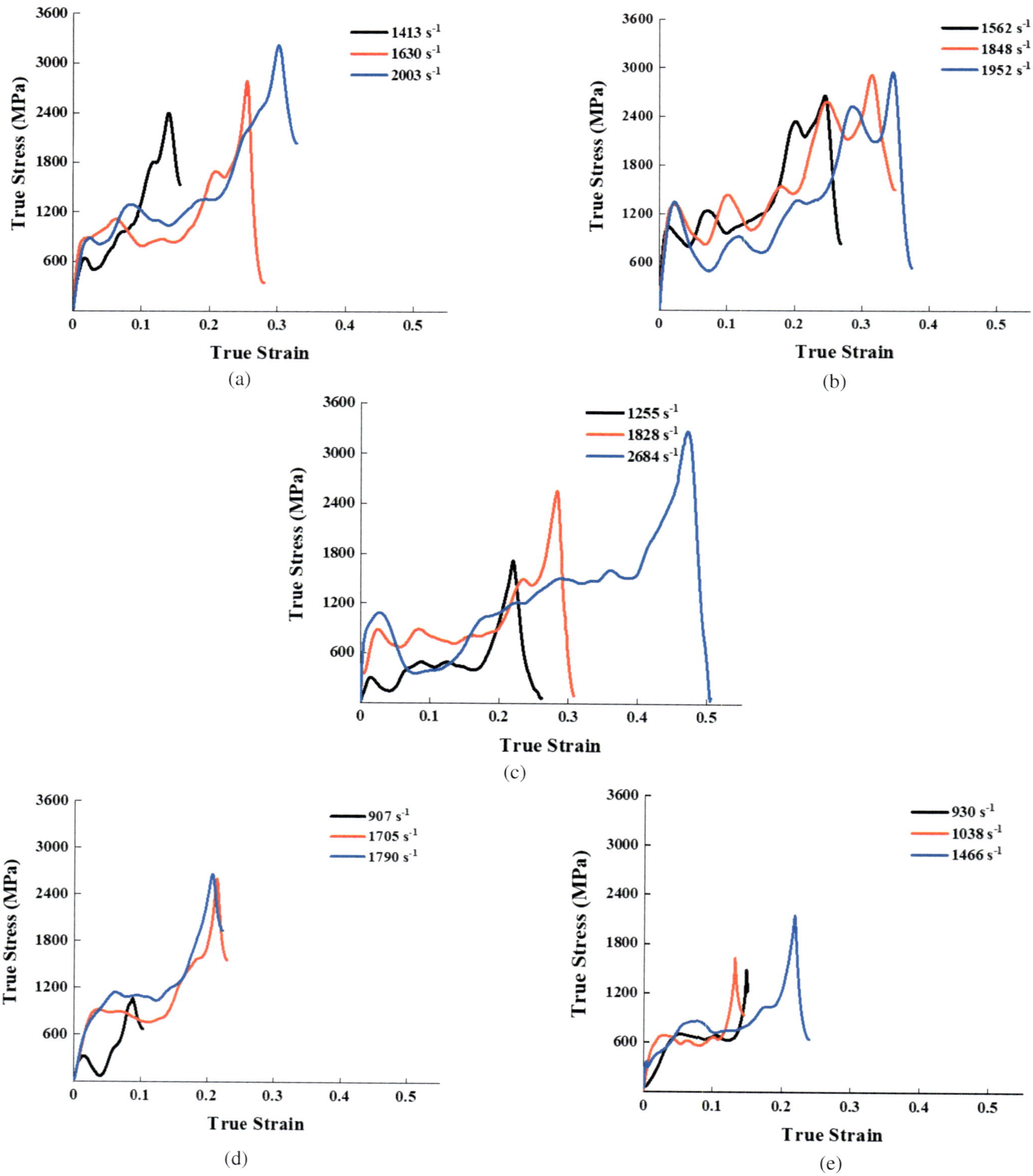

Fig. 3 True stress versus true strain plots for carbon/epoxy composite at various temperatures for different strain rates

Table 1 Mechanical properties of plain weave carbon/epoxy composite at various strain rates and temperatures

Temperature (°C)	Strain Rate (s^{-1})	Strength (MPa)	Strain at Peak stress
	2003	3219.057±3.121	0.303
- 45	1630	2781.013±5.431	0.255
	1413	2397.694±5.565	0.139
	1952	2951.927±1.233	0.346
- 25	1848	2917.760±4.632	0.315
	1562	2658.971±3.101	0.245
	2684	3269.564±3.060	0.471
25	1828	2556.910±3.557	0.283
	1255	1719.870±2.110	0.22
	1790	2661.967±6.812	0.207
35	1705	2604.910±5.205	0.214
	907	1055.401±4.273	0.087
	1466	2153.979±8.615	0.218
45	1038	1631.597±6.202	0.132
	930	1484.318±4.931	0.148

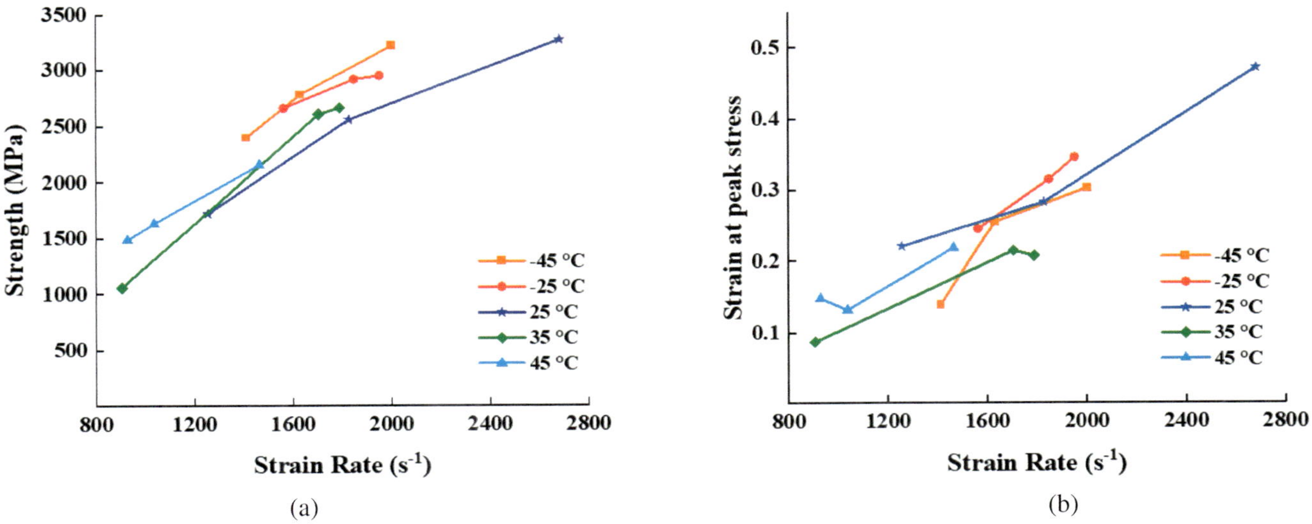

Fig. 4 Variation of (a) strength and (b) strain at peak stress of plain weave carbon/epoxy composite with strain rate at different temperatures

that with decreasing temperature, the strength of the carbon/epoxy composite increases which can be attributed to matrix stiffening and strain rate hardening. At lower temperatures, the epoxy matrix stiffens, allowing more stress to be transferred to the carbon fibers and decreasing fiber pullout. From fig. 3, it can be observed that at elevated temperatures, the matrix demonstrates enhanced ductility, facilitating improved stress transfer among fibers, resulting in a more gradual increase in stress and greater overall strain prior to failure.

The stress-strain curve also displays presence of some fluctuations before reaching ultimate stress. These fluctuations reduce when the composite specimen is subjected to higher temperatures. The presence of such fluctuations indicates that the composite laminate experienced initiation and progression of damage and cracks [15]. With decrease in temperature, the epoxy matrix becomes more brittle leading to matrix cracking. These cracks induce abrupt stress drops and generates stress wave reflections, which thereby introduces fluctuations in the stress -strain curve. The stress is further increased since the load is redistributed to the area surrounding the failed region [16]. The ductility of matrix and energy dissipation are enhanced at elevated temperatures, which thus reduces the severity of these fluctuations.

The images of fractured specimens are depicted in fig. 5 corresponding to temperature of 45 °C and – 45 °C. Figure 5 (a) depicts the deformed specimen when subjected to a temperature of 45°C at high strain rate of 1466 s^{-1}. The figure shows some fiber pull-out along with stepped pattern indicating interlaminar shear failure. This can be attributed to the fact that at moderate temperatures, epoxy softens slightly, reducing brittleness. Furthermore, fig. 5 (b) clearly shows that at cryogenic temperature of – 45 °C and strain rate of 1630 s^{-1}, carbon/epoxy composite experiences fibre breakage indicating characteristics of brittle fracture. This is further confirmed from fig. 3 (a), that demonstrates a sudden drop in stress following the peak stress. Thus, it can be concluded, that with increasing temperature, the carbon/epoxy composite experiences a transition in its failure mode from brittle fiber breakage to a failure mode that is a combination of shear failure and fiber pull out. This is also confirmed from the stress strain curves in fig. 3, with increase in failure strain and more gradual stress drop after peak stress.

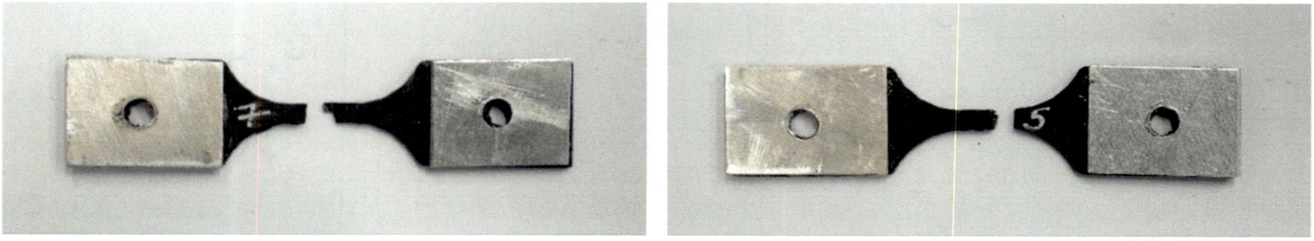

(a) (b)

Fig. 5 Deformed specimens when subjected to (a) $\dot{\varepsilon} = 1466$ s^{-1} at 45 °C and (b) $\dot{\varepsilon} = 1630$ s^{-1} at -45 °C

Conclusion

The current study focusses on the effect of temperature and high strain rate under tensile loading conditions on plain weave carbon/ epoxy composite laminates. The high strain rate experiments were conducted in the range of 900-2700 s^{-1} at temperatures of – 45 °C, – 25 °C, 25 °C, 35 °C and 45 °C utilizing tensile split Hopkinson pressure bar. Following are the conclusions drawn from the current study:

1. The strength and strain at peak stress of carbon/epoxy composite increases with increasing strain rate, indicating its strain rate sensitivity.
2. The decrease in temperature leads to increase in strength of carbon/epoxy composite, indicating thermal sensitivity of carbon/epoxy composite laminate. The strength increased by approximately 11% and decreased by 30%, when the temperature was varied to – 45 °C and 45 °C respectively, from room temperature (25 °C).
3. With decrease in temperature, matrix cracking occurs in the carbon/ epoxy composite due to brittleness of epoxy, leading to fluctuations in stress before attaining ultimate strength.
4. The variation in temperature influences the failure mode of the laminate.
5. With decrease in temperature, the failure mode of carbon/epoxy composite undergoes transition from fiber pull-out and shear failure to fiber rupture.

References

1. Koerber H, Xavier J, Camanho PP, "High strain rate characterisation of unidirectional carbon-epoxy IM7-8552 in transverse compression and in-plane shear using digital image correlation", Mechanics of Materials. 42(11),1004-19, 2010.
2. Academic Block. Image Based Measurement and Analysis, https://www.academicblock.com , 2023.
3. Woo SC, Kim TW. "High strain-rate failure in carbon/Kevlar hybrid woven composites via a novel SHPB-AE coupled test", Composites Part B: Engineering, 97,317-28, 2016.
4. Hosur M V, Adya M, Jeelani S, Vaidya UK, Dutta P K, "Experimental Studies on the High Strain Rate Compression Response of Woven Graphite/Epoxy Composites at Room and Elevated Temperatures", Journal of Reinforced plastics and composites, 23(5), 491-514, 2004.
5. Zhang X, Zhu D, Yao Y, Zhang H, Mobasher B, Huang L, "Experimental study of tensile behaviour of AFRP under different strain rates and temperatures", Journal of Structural Integrity and Maintenance, 1:1, 22-34, 2016.
6. Nakai K, Fukushima T, Yokoyama T, Arakawa K, "High strain-rate compressive properties of carbon/epoxy laminated composites–Effects of loading direction and temperature", In EPJ Web of Conferences ,183, 02011, EDP Sciences, 2018.
7. Jia S, Wang F, Zhou J, Jiang Z, Xu B, "Study on the mechanical performances of carbon fiber/epoxy composite material subjected to dynamical compression and high temperature loads", Composite Structures, 258, 113421, 2021.

8. Reis J M L, Coelho J L V, Monteiro A H, da Costa Mattos H. S, "Tensile behavior of glass/epoxy laminates at varying strain rates and temperatures", Composites Part B: Engineering, 43(4), 2041-2046, 2012.
9. Zhang HA, Yao YM, Zhu DJ, Mobasher B, Huang L, "Tensile mechanical properties of basalt fiber reinforced polymer composite under varying strain rates and temperatures", Polymer Testing, 51:29–39, 2016.
10. Joshi, P, Panigrahi S, "Evaluation of Tensile properties of FRP Composite Laminates under Varying Strain Rates and Temperatures", Fracture and Structural Integrity, 16(61), 338-351, 2022.
11. Chakraborty P, Singha M K, Tiwari V, "Dynamic tensile and compressive behaviour of AA7475-T7351 alloy under different strain rates and temperatures", Structures ,46, 49-63, Elsevier, 2022.
12. Pandouria A K, Yadav K, Tiwari V, "Compressive and tensile behavior of AA2014-T6 under different strain rates and different temperatures", Structures, 49,12-25, Elsevier, 2023.
13. Chen W W, Song B, "Split Hopkinson (Kolsky) bar: design, testing and applications", Springer Science & Business Media, 2010.
14. Kumar R, Bhagoria P, Bharadwaj M R, Tiwari V, "From quasi-static to dynamic: Experimental study of mechanical and fracture behaviour of epoxy resin", International Journal of Impact Engineering, 195, 105101, 2025.
15. Zhang W, Pan Z, Gideon RK, Gu B, Sun B, "Effects of temperature and strain rate on impact compression behaviors of three-dimensional carbon fiber/epoxy braided composites", Journal of Composite Materials, 49(7),771-82, 2015.
16. Hosur MV, Islam SW, Vaidya UK, Dutta PK, Jeelani S, "Experimental studies on the punch shear characterization of satin weave graphite/epoxy composites at room and elevated temperatures", Materials Science and Engineering: A. 368(1-2),269-79, 2004.

Chapter 6
Influence of Geometry and Loading Direction on the High Strain Rate Compression Response of modified GLARE Laminates

Muddu Rahul Bharadwaj*, **Rohit Kumar, Palak Bhagoria, and Vikrant Tiwari**

Abstract Fiber Metal Laminates (FMLs) are utilized across various applications because of their important benefits, which include high impact resistance, low density, and high strength to weight ratio. In this work, the compression behavior of plain-woven S-glass fiber-reinforced laminate interspersed with AA8011-H14 aluminum sheets (modified GLARE) under high strain rate loading is investigated using Split Hopkinson Pressure Bar (SHPB). The FML was produced using the vacuum-assisted resin transfer method. Under out-of-plane compression, two distinct specimen geometries—cuboidal and cylindrical— were examined at various strain rates ranging from 300 to 2000 s^{-1}. Additionally, rectangular specimens were also used to assess FML's dynamic compressive response in the in-plane direction. The results showed an increase in the compressive strength with strain rate for both the specimen geometries, yet in comparison, the cylindrical specimens exhibited a slightly higher strength to their rectangular counterparts at high strain rates due to better stress distribution. Delamination and fiber matrix cracking were the main failure modes for both the geometries. However, even at low strain rates, in cuboidal specimens localized failures, such as edge shear and corner cracking were observed. For the in-plane loading of rectangular specimens, metal layer buckling and plastic deformation dominated at lower strain rates, whereas brittle fracture of fibers and cracking in aluminum layers were seen at higher strain rates for the in-plane configuration.

Keywords FiberMetalLaminate · GLARE · SHPB · High strain rate

Introduction

Fiber Metal Laminates (FMLs) are a promising class of hybrid composite materials that offer the unique combination of properties of both metals and composites. The earliest evidence of FMLs can be dated back to the 1980s at the Delft university with the development of Aramid fiber reinforced Laminate (ARALL) which is an aluminium laminate reinforced with aramid fibers [1, 2]. Further development in this field with various FMLs made from other materials [3, 4] led to the production of Glass Fiber Reinforced Aluminum Laminates (GLARE). GLARE panels find numerous applications in aerospace, defense and marine industries due to their specific weight to strength ratio and their better adhesion properties compared to ARALL [5]. Due to these favorable properties, various researchers have studied the characterization of GLARE laminates subjected to uni-axial loading conditions [6, 7]. Most of the studies took help of the properties of individual constituents like Glass Fiber Reinforced Plastic (GFRP) [8], thermosetting or thermoplastic polymer used for adhesion [9] and the metal layers. Techniques like Digital Image Correlation (DIC) have also been employed in experiments along with numerical analysis [10, 11]. The dynamic characterization of these is also very crucial due to their nature of application. However, most experiments in the dynamic strain rate range focused on the tensile and impact resistance of the GLARE laminates [12–14]. For instance, Sharma et al. [12] studied the behavior of GLARE laminates with modifications to their thickness under high strain rate tensile loading. The FMLs showed strain rate sensitive behavior at high strain rates and the strength was shown to increase by 60% compared to the quasi-static loading conditions. Zhou et al. [13] along with high strain rate tensile experiments, have extended their study to low velocity impact tests on specimens made from modified GLARE by using different metal alloys. Their findings showed a contrast of low-strain rate sensitivity for the range of strain rates considered.

Muddu Rahul Bharadwaj · Rohit Kumar · Palak Bhagoria · Vikrant Tiwari
Department of Applied Mechanics, Indian Institute of Technology Delhi, New Delhi, India, 110016
e-mail: rahulbharadwaj95@gmail.com; rohitnituk001@gmail.com; palakbhagoria@gmail.com; Vikrant.Tiwari@am.iitd.ac.in

Paul Allison et al. (eds.), *Dynamic Behavior of Materials, Vol. 1*, Conference Proceedings of the Society for Experimental Mechanics Series, https://doi.org/10.13052/9788743808275-6

In this study, efforts have been made to study the dynamic behavior of modified GLARE laminate which is made from plain-woven S-glass fiber reinforcement and AA8011-H14 aluminium sheets. The geometrical effect of the typical cylindrical vs cuboidal specimens was investigated at high strain rates of 300 – 2000 s^{-1}. Additionally, these cuboidal specimens were also subjected to in-plane loading conditions along the thickness direction and the failure types were observed and reported.

Methodology

Split Hopkinson Pressure Bar (SHPB) or Kolsky bar is a commonly used technique to evaluate the material's response at high strain rates. The SHPB used in this study consists of three bars, namely, the incident, transmission and striker bar, all made of titanium (Ti-6Al-4V) material. Working principle of the SHPB is very well documented by many researchers [15, 16]. The strain vs time signals obtained from the data acquisition system of the SHPB setup are further used to calculate the average stress (σ_s) – strain (ε_s) relation and the strain rate ($\dot{\varepsilon}$) of the specimen using the following equations [17].

$$\dot{\varepsilon} = \frac{2C_b}{L_s}\varepsilon_r \tag{1}$$

$$\sigma_s(t) = \frac{A_b E_b}{A_s}\varepsilon_t(t) \tag{2}$$

$$\varepsilon_s = \frac{2C_b}{L_s}\int_0^t [\varepsilon_r(t)]\, dt \tag{3}$$

Here, C_b, E_b, A_b, represent the wave velocity, Elastic modulus and area of the bar respectively. A_s, L_s are the area and length of the specimen respectively. $\varepsilon_{r,}\varepsilon_t$are the reflected and transmitted strain signals and *t* being the time. A typical strain signal obtained from the SHPB experiments is shown below (Fig. 1) along with the force equilibrium that is achieved over both ends of the specimen.

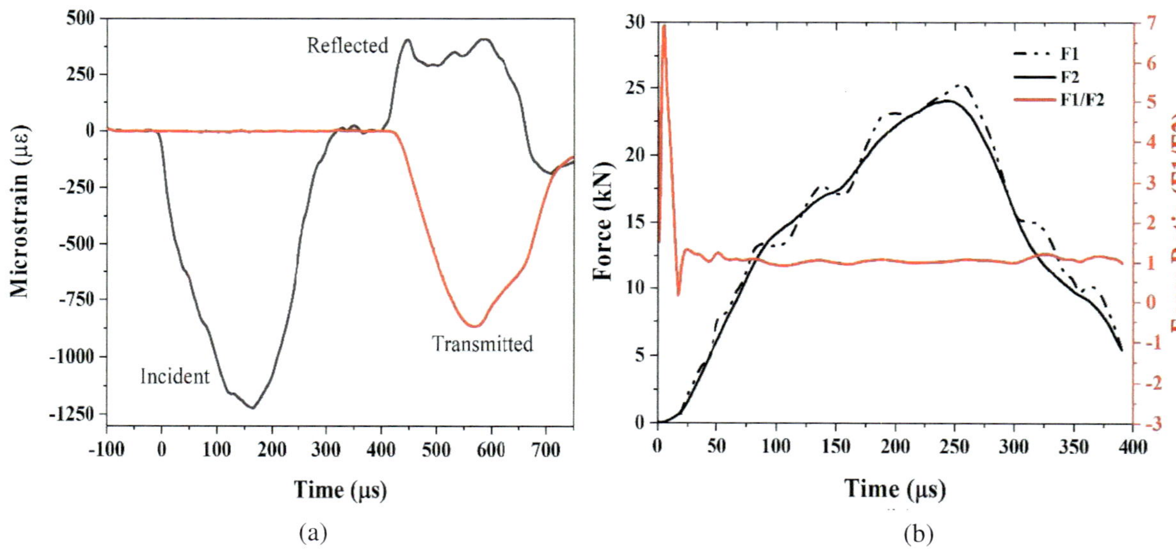

Fig. 1 (**a**) Strain – time history and (**b**) force equilibrium of a typical experiment

The FML utilized for this study was prepared using Vacuum Assisted Resin Transfer Molding (VARTM) method. It consists of 4 layers of AA8011-H14 sheets of thickness 0.5 mm each and 3 composite layers (Fig. 2). Each composite layer is made up of five plain woven S-glass fabric sheets bonded with two-part epoxy resin system (EPOFINE® –1564/FINEHARD-3486–2) supplied by Fine Finish Organics Pvt. Ltd, Mumbai. The metal volume fraction was 0.6, while the remaining 0.4 fraction, which included fiber and epoxy, had the fiber volume fraction of 0.6. The final dimensions of the FML so produced were 300 mm by 300 mm by 6 mm. The samples required for dynamic compression were prepared using high pressure waterjet cutting. The dimensions of cylindrical and cuboidal specimens so produced were 12 mm (diameter,

Fig. 2 Cuboidal and cylindrical specimens used in the study showing the layup sequence

D) by 6 mm (thickness, L) and 11 mm by 11 mm by 6 mm respectively. The L/D ratio for out-of-plane experiments was maintained to be 0.5.

Results and Discussion

Out-of-Plane Loading

The out-of-plane compression experiments were done on SHPB for both cylindrical and cuboidal specimens at various strain rates ranging from 300 – 2000 s^{-1}. For ease of comparison and data analysis, the strain rates were divided into low, medium and high strain rate regimes within the range: 300 – 900 s^{-1}, 900 – 1500 s^{-1}, 1500 – 2000 s^{-1} respectively. The stress – strain response of cylindrical and cuboidal specimens for three different strain rates is shown in Fig. 3.

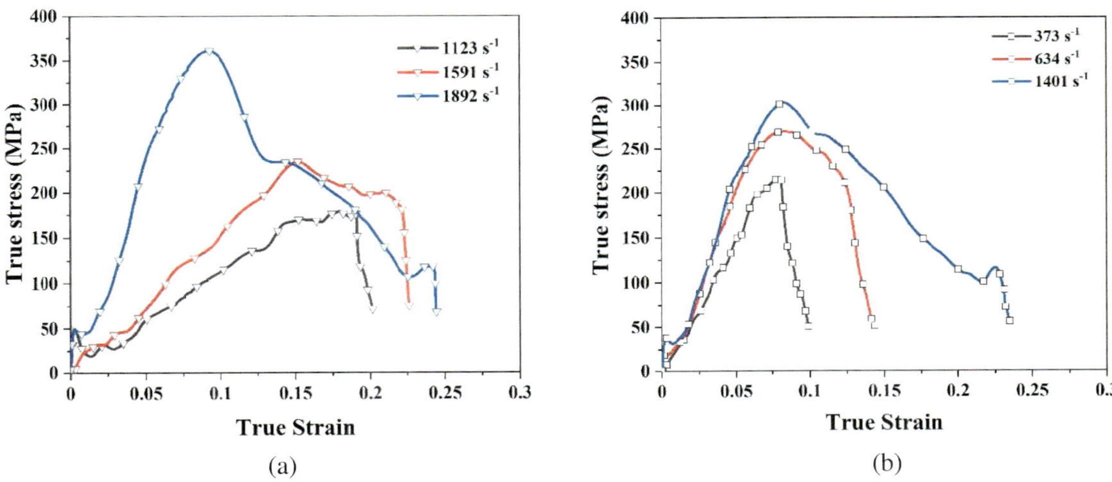

Fig. 3 Stress – strain response of (**a**) cylindrical and (**b**) cuboidal geometry specimens at various strain rates

Both geometries showed positive strain rate sensitivity. For cylindrical specimens, in the initial linear elastic region, the slope remained almost constant indicating stiffness is relatively insensitive to strain rate. Also, the final failure is different at higher strain rate regime with a more pronounced dip indicating brittle failure compared to the more progressive failure at low and mid strain rate regimes. For the cuboidal specimens, a similar trend of stiffness insensitivity to strain rate was observed. However, the initial slope was a little steeper compared to their counterparts. The post yield hardening response also was more rapid and pronounced possibly due to non-axisymmetric shape. Nevertheless, the primary reason for failure was found to be delamination and fiber matrix cracking for both the geometries. A power law model, as shown in Eq.4, was used to find out the relation between ultimate strength (σ_{ult}) and strain rate ($\dot{\varepsilon}$) for both the geometries.

$$\sigma_{ult} = C\dot{\varepsilon}^{\beta}. \tag{4}$$

C is the proportionality constant and β is the strain rate sensitivity parameter. The ln σ_{ult}- ln $\dot{\varepsilon}$ plots for both the geometries can be seen in Fig. 4. As exhibited earlier, both geometries have positive strain rate sensitivities. However, the cylindrical specimens have shown higher strain rate sensitivity than the cuboidal ones. All these parameters for the linear curve fit are shown in Table 1.

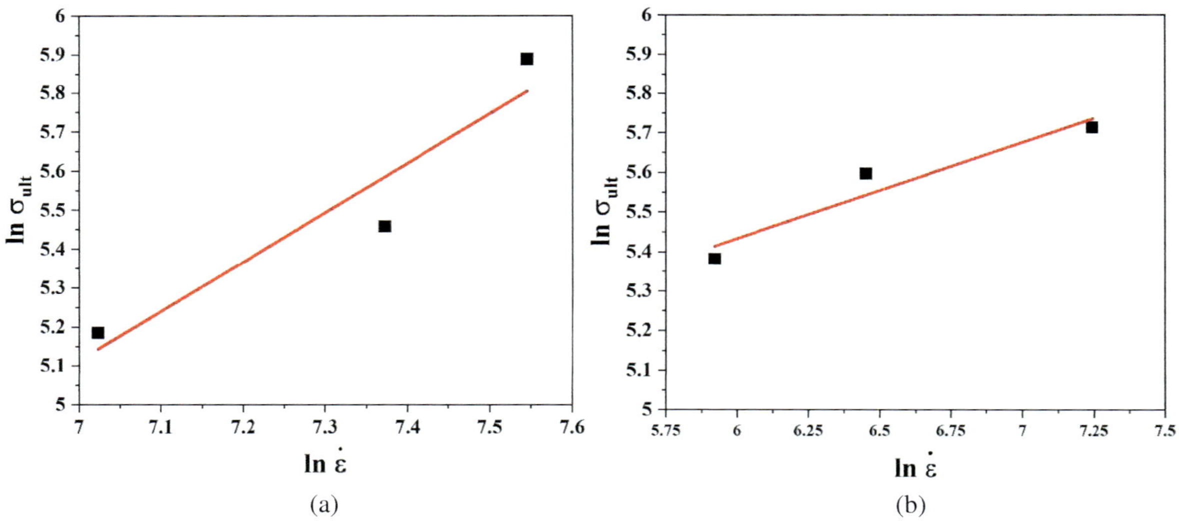

Fig. 4 ln-ln plots of stress vs strain rate for (**a**) cylindrical and (**b**) cuboidal specimens along with a linear fit

Table 1 Parameters of curve fitting used in Fig. 4 for cylindrical and cuboidal specimens

	Equation: y = a + b*x	
	Cylindrical	***Cuboidal***
Intercept (a)	-3.76214 ± 3.06711	3.97603 ± 0.46552
Slope (b)	**1.26787 ± 0.41918**	**0.24282 ± 0.07094**
R-Square (COD)	**0.90146**	**0.92136**
Residual Sum of Squares	0.0248	0.00446

Using the above linear fit parameters, the ultimate stresses for the missing strain rates in various regimes were extrapolated and found out. A new dimensionless parameter called strength ratio (σ_o^*), defined as per the following equation to facilitate comparison, is used.

$$\sigma_o^* = \frac{\sigma_{cylindrical}}{\sigma_{cuboidal}}. \quad (5)$$

It has been found that the cuboidal specimens generally reached the higher peak stress values at lower strain rates, indicating their earlier failures and stress concentrations at edges and corner cracking. This parameter has been plotted against the strain rate for a better understanding of its sensitivity (Fig. 5). As the strain rate increased, the value of strength ratio which was less than one slowly rose to higher than 1 which shows the sustained strain hardening effects in cylindrical specimens. A polynomial fit was determined to best describe its variation indicating a non-linear sensitivity to strain rate. The parameters of the curve fit are given in Table 2.

Table 2 Polynomial curve fit parameters of strength ratio from Fig. 5

Equation	y = C1 + B1*x^1 + B2*x^2
Intercept (C1)	**0.09808 ± 0.14398**
B1	**2.75051E-4 ± 2.97016E-4**
B2	**1.17312E-7 ± 1.31222E-7**
R-Square (COD)	**0.969**
Residual Sum of Squares	**0.01554**

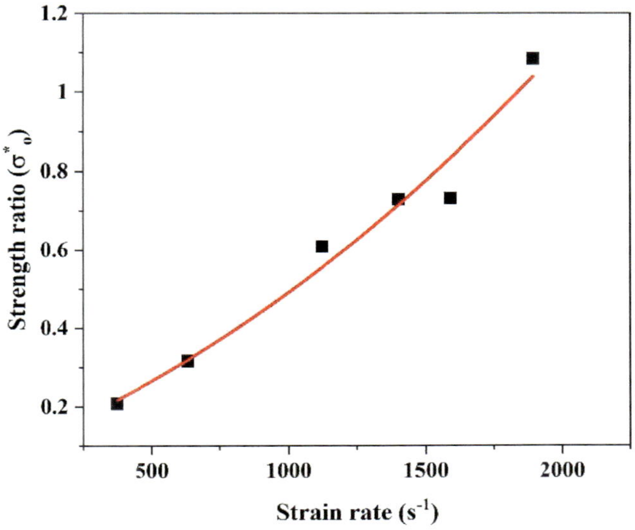

Fig. 5 Strength ratio vs strain rate with a polynomial fit.

In-Plane loading

The cuboidal specimens were also loaded along the warp direction (in-plane loading) i.e. along the thickness direction to study the effect of dynamic compression at various strain rates ranging from 600 – 1500 s^{-1}. Although the effective area is reduced, the volume of the material remains constant. Hence, an effort has been made to study their behavior and compare them with out-of-plane cuboidal specimens. The stress – strain plots of the cuboidal specimens loaded along the in-plane direction are shown in Fig. 6 (a). A similar attempt to find out the strain rate sensitivity of in-plane loaded is made using the Eq. 4 and the ln σ_{ult}- ln $\dot{\varepsilon}$ is plotted (Fig. 6 (b)). Through a linear fit, the strain rate sensitivity parameter (β) is found out and compared to that of the out-of-plane loaded specimens (Table 3). Buckling/bending of aluminum layers at lower strain rates was observed as the primary failure mechanism during this loading. The strain-to-failure and the peak stress achieved through this loading are comparatively lower, indicating early failures of aluminum due to its thickness. Furthermore, higher localization of deformation and cracking was observed within the metal parts of the FML, suggesting their instability, along with fiber cracking as the strain rate increased.

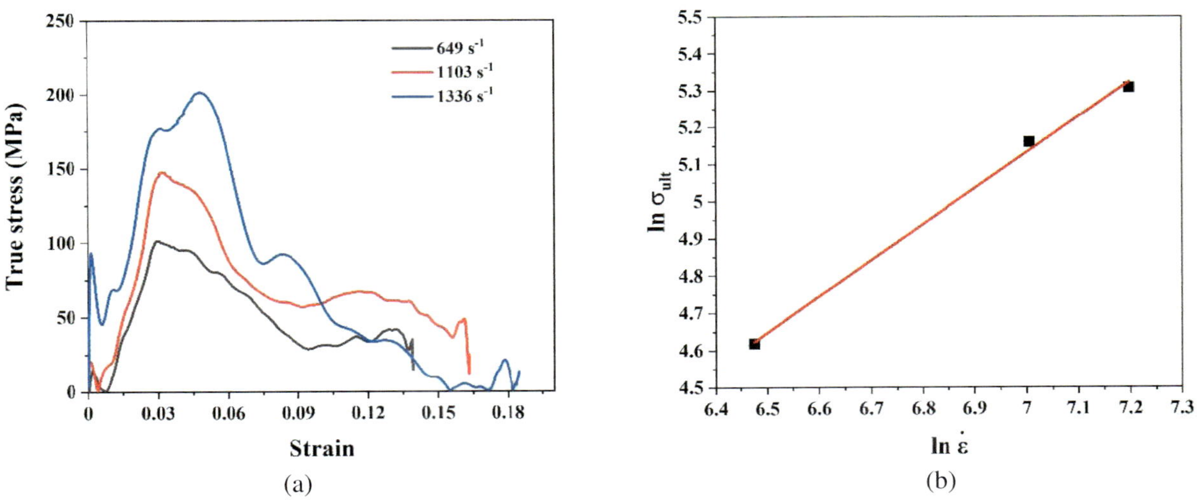

Fig. 6 (**a**)Stress - strain response and (**b**) stress - strain rate behavior of Cuboidal samples in in-plane loading

The strain rate sensitivity is higher than the out-of-plane loading conditions as can also be observed from the stress-strain plots which show a very steep increase in the peak stress values.

Table 3 Linear curve fit parameter from Fig. 5

Equation	y = a + b*x
Intercept	**-1.64263 ± 0.36314**
Slope	**0.96771 ± 0.05263**
R-Square (COD)	**0.99705**
Residual Sum of Squares	7.74975E-4

CONCLUSION

This study was aimed at finding out the effect of geometry and loading direction of high strain rate compression response of AA8011/ Plain woven S-Glass fiber epoxy modified GLARE laminates. The experiments were carried out in the range of 300 – 2000 s^{-1} initially for both cylindrical and cuboidal specimens along ot-of-plane loading direction. The strain sensitivity parameter was found out using power law model with the best fit being linear. At lower strain rates, although cuboidal specimens had higher peak stress values, with increase in strain rate this trend was reversed, indicating their very low strain rate sensitivity. The damage too was more progressive in case of cylindrical specimens with stress uniformity compared to the more brittle nature and steep rise in plastic zones observed in cuboidal specimens due to their non-axisymmetry. The major damage mechanism was the delamination of the metal and composite layers in both the geometries. Further, in-plane loading experiments were also conducted on the cuboidal samples along the warp (thickness direction) for strain rates of 600 – 1500 s^{-1}. The strain rate sensitivity was on par with the out-of-plane cylindrical experiments except the strain to failure and peak stress values being very low. Buckling of the aluminum layer and fiber fracture were the dominating failure mechanisms in this direction of loading.

References

1. Vermeeren CA, "An historic overview of the development of fibre metal laminates", *Applied Composite Materials*,10:189-205, 2003.
2. Vlot A, Gunnink JW, "Fibre metal laminates: an introduction", *Springer science & Business Media*, 2011.
3. Asundi A, Choi AY,"Fiber metal laminates: an advanced material for future aircraft", *Journal of Materials processing technology*, 63(1-3):384-94, 1997.
4. Cantwell WJ, "The mechanical properties of fibre-metal laminates based on glass fibre reinforced polypropylene", *Composites science and technology*, 60(7):1085-94, 2000.
5. Young JB, Landry JG, Cavoulacos VN, "Crack growth and residual strength characteristics of two grades of glass-reinforced aluminium 'Glare'", *Composite Structures*, 27(4):457-69, 1994.
6. Sharma AP, Khan SH, Parameswaran V, "Experimental and numerical investigation on the uni-axial tensile response and failure of fiber metal laminates". *Composites Part B: Engineering*. 125:259-74, 2017.
7. Ergun H, Liaw BM, Delale F, "Experimental-theoretical predictions of stress–strain curves of glare fiber metal laminates", *Journal of Composite Materials*, 52(1):109-21, 2018.
8. Song B, Chen W, Weerasooriya T, "Quasi-static and dynamic compressive behaviors of a S-2 glass/SC15 composite", *Journal of Composite Materials*,37(19):1723-43, 2003.
9. Kumar R, Bhagoria P, Bharadwaj MR, Tiwari V, "From quasi-static to dynamic: Experimental study of mechanical and fracture behaviour of epoxy resin", *International Journal of Impact Engineering*.195:105101, 2025.
10. Academic Block, "Image Based Measurement and Analysis", *https://www.academicblock.com*, 2023.
11. He W, Wang C, Wang S, Yao L, Wang L, Xie D, "Characterizing and predicting the tensile mechanical behavior and failure mechanisms of notched FMLs—Combined with DIC and numerical techniques", *Composite Structures*, 254:112893, 2020.
12. Sharma AP, Khan SH, Parameswaran V, "Response and failure of fiber metal laminates subjected to high strain rate tensile loading", *Journal of Composite Materials*, 53(11):1489-506, 2019.
13. Zhou J, Guan ZW, Cantwell WJ, "The influence of strain-rate on the perforation resistance of fiber metal laminates", *Composite structures*,125:247-55, 2015.
14. Khan SH, Sharma AP, Kitey R, Parameswaran V, "Effect of metal layer placement on the damage and energy absorption mechanisms in aluminium/glass fibre laminates", *International Journal of Impact Engineering*, 119:14-25, 2018.
15. Chakraborty P, Pandouria AK, Singha MK, Tiwari V, "Constitutive behavior of AA7475-T7351 at high strain rate and elevated temperatures", *Dynamic Behavior of Materials*, Volume 1: Proceedings of the 2020 Annual Conference on Experimental and Applied Mechanics 2021 (pp. 15-23), Springer International Publishing.
16. Pandouria AK, Yadav K, Tiwari V, "Compressive and tensile behavior of AA2014-T6 under different strain rates and different temperatures", *Structures*, Vol. 49, pp. 12-25, Elsevier. 2023.
17. Chen WW, Song B, "Split Hopkinson (Kolsky) bar: design, testing and applications,"*Springer Science & Business Media*, 2010.

Chapter 7
Wave Control in Hierarchical Kerf Structures

Samuel Crawford, Kanak Mantri, Mihir Patil, Marcia A. Cooper*, and Anastasia Muliana

Abstract The purpose of this research is to understand the energy storing, dissipation abilities, and stress wave propagation in three-dimensional kerf structures under dynamic impact loading conditions. We study a square kerf pattern made of multiple polymer materials. The first approach uses a single polymer of either brittle polylactic acid (PLA) or compliant thermoplastic polyurethane (TPU). The second approach uses a programmable filament (PF) having a mixture of PLA and TPU to create composite microstructures with alternating brittle and compliant materials. The deformation responses indicated that the cell reconfigurations, from buckling and densification, and cell fracture during loading contribute to energy dissipation while energy storing is attributed to the material elastic responses. We expect that by tuning kerf geometrical parameters and altering materials within the kerf structures, we can control wave propagation and energy management. To understand the dynamic response and deformation patterns, the kerf structures are subjected to intermediate impact from a split Hopkinson pressure bar (SHPB) facility. During testing, kerf structure deformation is recorded at high speed and strain gauge signals are recorded. The experimental deformation patterns and stress wave propagation responses are simulated using ABAQUS FEA. We first establish material models for the 3D printed materials at high strain rates, demonstrating the ability to simulate their material responses in dynamic compression, and then explore the energy dissipation and stress wave propagation in 3D hierarchical kerf structures. Kerf structures are potentially used for controlling energy transfer under dynamics loadings.

Keywords Kerf Structure · Dynamic Impact · Wave Propagation

Introduction

Recent advancements in manufacturing techniques and technologies allow for the creation of complex geometrical structures that exhibit improved mechanical properties beyond the base material. In this study, we adopt a kerfing topology, traditionally used in wood cutting, to create thick and flexible structures that are reconfigurable to complex geometries [1]. Studies in planar and thin-walled kerf structures show that the structures can be used to control wave propagation and energy dissipation through the structure [2, 3]. These properties of kerf structures make them potentially beneficial in engineering applications where energy management is required in a dynamic, high-stress environment, such as in aerospace or defense [4, 5]. By utilizing 3D printing technology, it is possible to create kerf structures additively using a range of polymers with widely varying material properties. Some studies have already explored the energy absorption capabilities of 3D printed lattice and cell structures at low strain rates [6–8].

In high strain rate environments, kerf structures may exhibit unique energy dissipation capabilities which are largely unstudied. The split-Hopkinson pressure bar apparatus is a standard method for material characterization at high strain rates [9, 10]. In this study, we characterize the stress-strain behavior of a 3D printed kerf structure made of several different polymer materials, including PLA, TPU, and PF based on combinations of 0.5 PLA and 0.5 TPU by volume [11].

Samuel Crawford · Kanak Mantri · Mihir Patil · Marcia A. Cooper · Anastasia Muliana
J. Mike Walker '66 Department of Mechanical Engineering
e-mail: spcrawford@tamu.edu; kanakmantri@tamu.edu; mihirpatil@tamu.edu; macoope@tamu.edu; amuliana@tamu.edu

Marcia A. Cooper
Materials Science and Engineering Department
Texas A&M University, College Station, Texas, USA 77843
e-mail: macoope@tamu.edu

Paul Allison et al. (eds.), *Dynamic Behavior of Materials, Vol. 1*, Conference Proceedings
of the Society for Experimental Mechanics Series, https://doi.org/10.13052/9788743808275-7

Materials and Methods

Two different structures were tested – short disks to establish material stress-strain behavior, and kerf structures to investigate energy management behaviors (Fig. 1). Each structure was 3D printed in PLA, TPU, or the custom blended PF. All samples were printed using a Bambu Lab P1P 3D printer with a 0.4 mm nozzle with a bed temperature of 60 °C, a layer height of 0.2 mm, and 100% in-fill. The PLA samples were printed using Overture 3D 1.75 mm diameter PLA filament with a printing temperature of 210 °C. TPU samples were printed using Overture 3D 1.75 mm diameter TPU filament with a printing temperature of 220 °C. PF samples were printed at 235 °C with a filament having a parallel arrangement of PLA and TPU with 50:50 volume ratio.

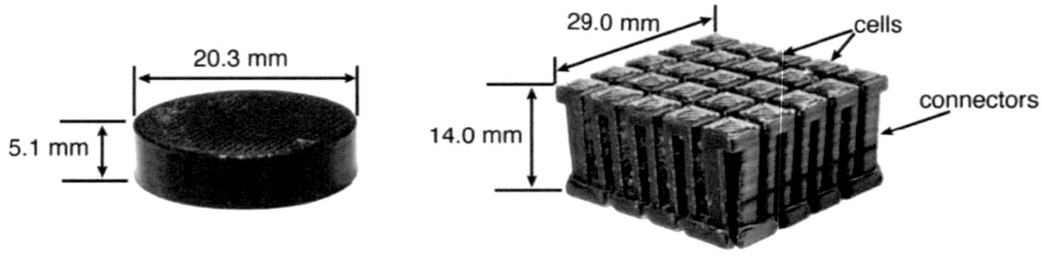

Fig. 1 Photographs of 3D printed disk and square kerf structure

The samples were tested in high strain rates with a conventional split Hopkinson pressure bar apparatus (SHPB, Fig. 2) consisting of a gas gun, striker bar, incident bar, transmission bar, and data measurement and acquisition system. In this standard test setup, the three bars are the same material and diameter. The sample is positioned in between and in contact with the incident and transmission bar. The gas gun launches the striker bar to impact the free end of the incident bar. The compressive pulse generated from this impact then travels through the incident bar until reaching the sample, where the impedance mismatch reflects a tensile pulse back through the incident bar and transmits a compression pulse through to the transmission bar. Our SHPB was fabricated by Dynamic Systems and Research, Inc. and utilizes a 25.4-mm diameter aluminum 7075-T6 compression bar set (incident bar length of 3.05 m and transmission bar length of 2.44 m). Striker bar velocity is measured with a Dynamic Systems & Research model DSR-1031 laser sensor.

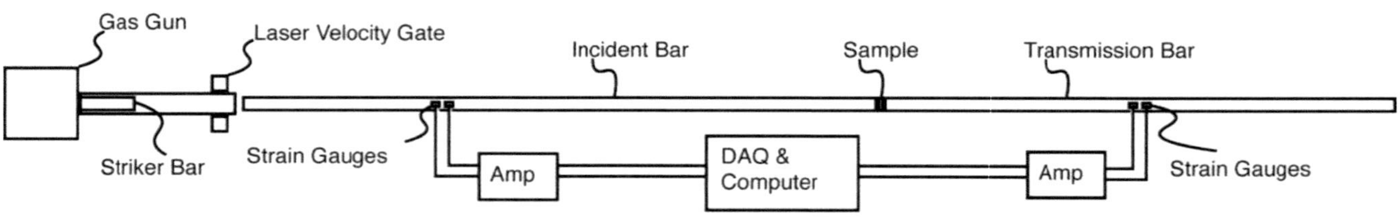

Fig. 2 Diagram of the split-Hopkinson pressure bar apparatus

Strain gauges on the incident and transmission bar measure the incident pulse strain (ε_i), reflected pulse strain (ε_r), and transmitted pulse strain (ε_t) histories. Two separate strain gauges each are attached to the incident and transmission bars – one Micro-Measurements EA-06-125BZA-350/E 350-Ohm strain gauge, and one Micro-Measurements WK-13-125BZ-10C 1000-Ohm strain gauge. Each strain gauge is connected to a Vishay Measurements Group Model 2310B signal conditioning amplifier. Signals from all the amplifiers and the velocity sensor are collected by a National Instruments cDAQ chassis and modules at a sampling rate of 2 MHz.

These measured strains allow for the calculation of sample stress (σ), strain rate ($\dot{\varepsilon}$), and strain (ε) by the following equations [9, 10]:

$$\sigma(t) = \frac{AE}{2A_s}(\varepsilon_i + \varepsilon_r + \varepsilon_t) \# (1)$$

$$\dot{\varepsilon}(t) = \frac{2c_b \varepsilon_r}{A_s} \# (2)$$

$$\varepsilon(t) = \int_0^t \dot{\varepsilon} d\tau \# (3)$$

where A is the cross-sectional area of the bars, E is the Young's modulus of the bars, A_s is the instantaneous cross-sectional area of the sample, and c_b is the bar sound speed. Thus, the stress-strain behavior of the sample material at the tested strain rate is calculated. This analysis is called a 3-wave analysis, for its use of the incident, reflected, and transmitted waves to calculate the sample stress [9].

This analysis relies on the assumption of force equilibrium at the sample faces in contact with the bars which occurs within a short transient for ductile materials and long compressive pulse durations. In compliant materials, achieving force equilibrium can be challenging and requires pulse shaping — the addition of a sacrificial, plastically deforming material to introduce a slope to the incident pulse that ramps the strain slowly — allowing the front and back stresses of the sample to remain in equilibrium for the duration of the pulse. To observe the stress-strain material behavior of the short disk samples, a series of different pulse shapers was used to achieve force equilibrium in the samples. For the PLA samples, a combination of copper and aluminum pulse shapers were used together as a dual pulse shaper. The aluminum and copper were both disks 8.95 mm (0.3525 inch) in diameter. The aluminum was 0.127 mm (0.005 inch) thick and placed against the incident bar free end. The copper was 0.8255 mm (0.0325 inch) thick and placed between the striker bar and the aluminum pulse shaper. In the case of TPU, a closed-cell neoprene foam was used. The neoprene pulse shapers were cylinders with a diameter and height each 12 mm. Due to limited number of samples, the PF samples were tested with the same pulse shaper used with the PLA and TPU tests.

As shown in Fig. 1, the size of the square kerf structure (29.0 mm × 29.0 mm) is greater than the SHPB bar diameter (25.4 mm) which was addressed by adding platens at the sample end of the incident and transmission bars for additional contact area (Fig. 3). The platens were attached to the bars via a threaded connection with thread relief to allow for a flat, parallel, intimate contact between the end of the bar and the inside face of the platen. The platens were designed and manufactured to maintain the standard best practices for SHPB interfacial flatness and parallelism. The diameter of the platen expands from the 25.4-mm bar diameter to a diameter of 63.5 mm (2.50 inch) over a length of 14.0 mm (0.550 inch).

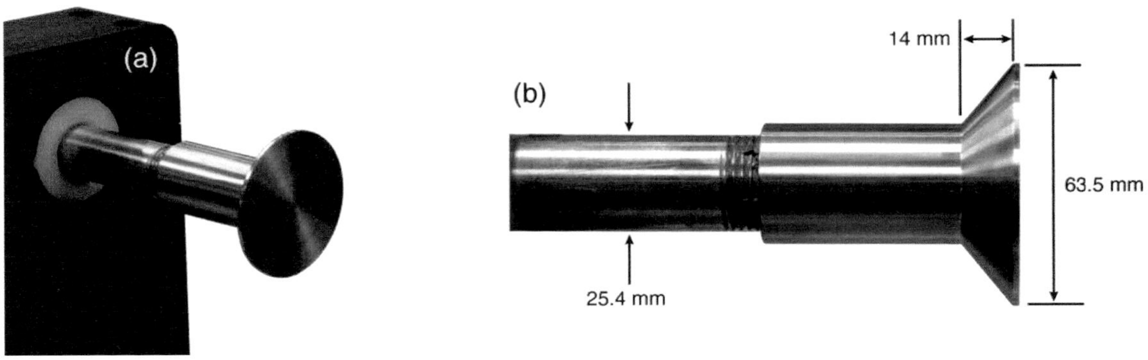

Fig. 3 Photographs of the incident bar with platen installed. An identical platen was also installed similarly on the transmission bar (not shown).

It is readily acknowledged that the area change of the platens causes the experimental method to deviate from the 1D assumptions of a traditional SHPB test. Bar-on-bar tests with the platens installed showed minor effect on the reflected and transmitted bar strains. Additionally, no pulse shaping was used when testing the kerf structures. As our motivation is to explore the effect kerf structure and material design, the platens enable large-scale structures to be explored while still enabling relative comparisons to be made. Complimentary FE simulations further extend our ability to attribute changes in energy management to material or structure effects. High speed video of the structures during dynamic compression were captured with a Photron SA-Z camera recording at 50 MHz.

Material characterization results

Experiments were performed to assess the dynamic stress-strain behavior of the three 3D-printed materials of the short disk samples. Fig. 4 (a) plots the stress-strain data for PLA under dynamic compression and modest strain rates between 170 - 490 s^{-1}. Repeat tests were performed under nominally similar impacts—striker velocity and pulse shaping—and average strain rates with standard deviations are reported with the data groupings in Fig. 4 (a). The brittle PLA transmitted most of the incident wave as observed in the strain data of Fig. 4 (b) where there is a minor difference between the incident (I) and transmitted (T) signals. At these low strain rates, evidence of damage was not visible on the external surfaces in post-test inspections. Although the PLA material under the above strain rates shows significant rate-dependent effect, it does not seem to dissipate energy.

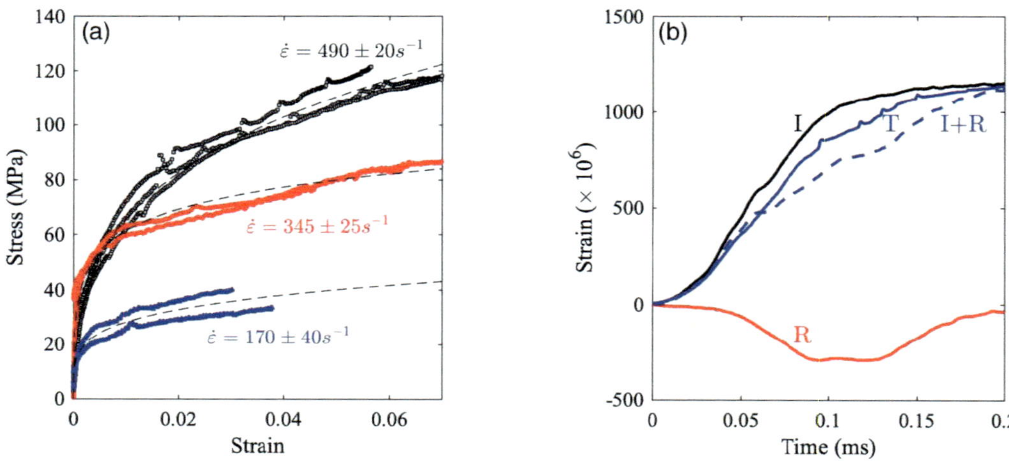

Fig. 4 (a) The strain histories of PLA under compression loading with reported average strain rate for each test. The reported strain rates are the average and standard deviation of near-replicate test conditions. (b) The strain signals reported from the I: incident, T: transmitted, R: reflected gauges for the test plotted in (a) at $\dot{\varepsilon} = 490$ s^{-1}

Fig. 5 (a) plots the stress-strain data for TPU under dynamic compression and higher strain rates between 1370 - 2190 s^{-1}. The compliant TPU deformed at much higher strain rates without evidence of damage. This material also reflected most of the incident wave as observed in the strain data of Fig. 5 (b) where there is a minor difference between the incident (I) and reflected (R) signals. The compliant TPU deformed significantly under dynamic compression absorbing the kinetic energy and storing them as elastic energy, allowing the TPU to regain its original shape after deformation.

Fig. 6 (a) plots the stress-strain data for PF under dynamic compression and higher strain rates between 1410 - 1810 s^{-1}. A repeat test was performed and the average strain rate with standard deviation is reported. It is interesting that the PF (a 50:50 volume mixture of PLA and TPU) exhibits a low sensitivity to strain rate, whereas independently PLA (Fig. 4 (a)) and TPU (Fig. 5 (a)) both showed a clear dependence on strain rate. The PF maximum strain and stress were more like TPU versus PLA as shown in Fig. 6 (b) for two tests with similar strain rate histories. Thus, mixture ratio does not define bulk behavior of these printed composite materials—as the PF stress-strain behavior is not simply an average of the individual PLA and TPU responses—but suggests that the composite material behavior may be 1) controlled by the behavior of the most compliant component or 2) strongly influenced by the interlayer bonding and adhesion of the printing process, or 3) some combination of these effects.

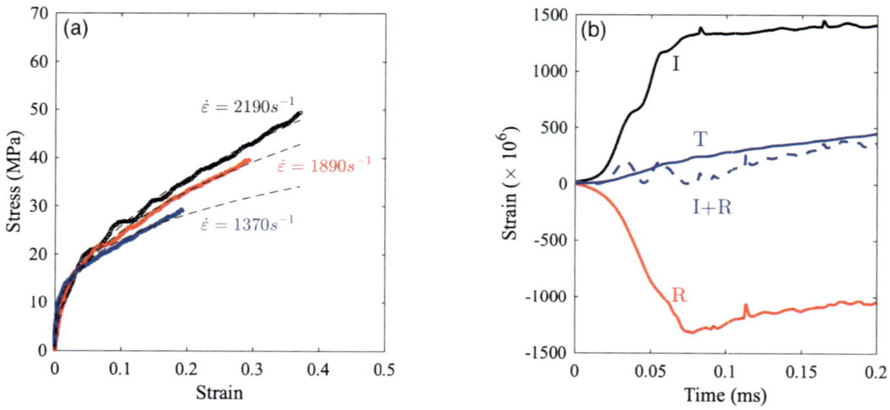

Fig. 5 (a) The strain histories of TPU under compression loading with reported average strain rate for each test. (b) The strain signals reported from the I: incident, T: transmitted, R: reflected gauges for the test plotted in (a) at $\dot{\varepsilon} = 2190$ s^{-1}

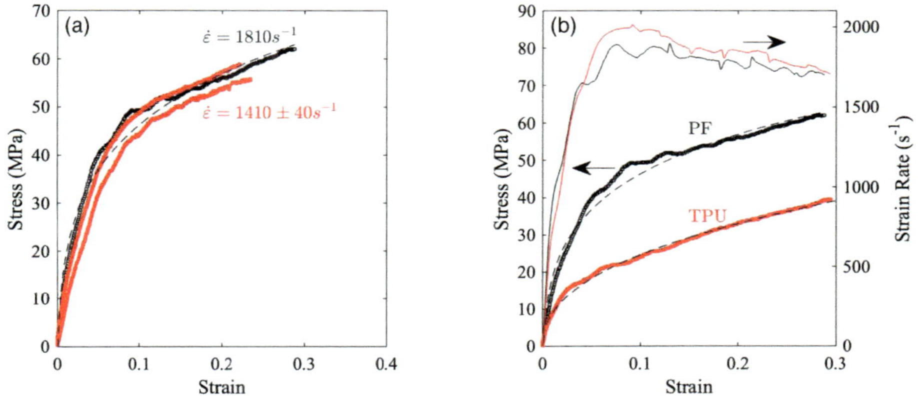

Fig. 6 (a) The strain histories for the PF under compression loading. The reported strain rates are the average and standard deviation of nominally similar test conditions. (b) A comparison of the TPU and PF stress-strain curves for the PF and TPU at similar strain rate conditions

Material model applied to square kerf structure

Using the data of Fig. 4 (a), Fig. 5 (a), Fig. 6 (a) for the three individual materials, simulations were performed on the short disk samples to demonstrate model capability. First, bar-on-bar simulations matched experimental elastic wave propagation including pulse durations and amplitudes. Then, simulated and experimental data for PLA are presented for PLA in Fig. 7 (a) for the short disk and in Fig. 7 (b) for the square kerf structure. We represented these materials with a Johnson-Cook rate-dependent material model, suitable for capturing the stress-strain responses shown above. The PLA disk transmits most of the incident stress wave, while the buckling/bending of kerf connectors alter the stress wave propagation, i.e., reducing transmission and increasing reflection. It is believed that deformations of kerf connectors enable the PLA kerf structure to partially absorb kinetic energy. Future study can optimize the geometrical parameters in PLA kerf structures to control wave propagation.

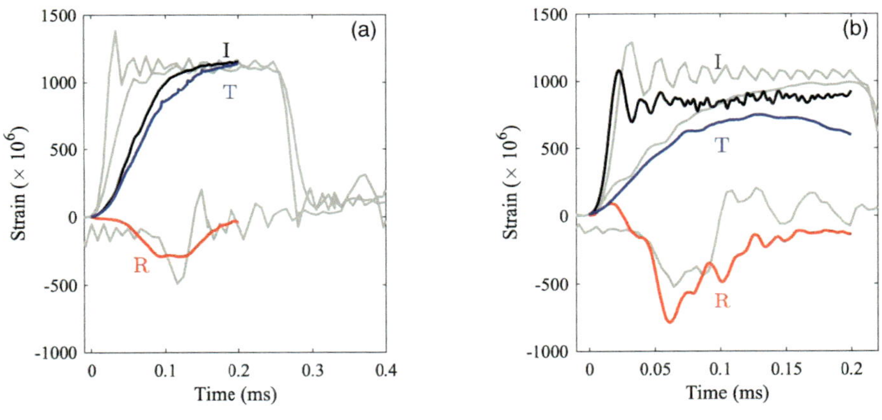

Fig. 7 Comparison of the experimental (solid-colored lines) and the simulated (gray lines) strain histories from the I: incident, T: transmitted, R: reflected gauges for (a) PLA disk sample and (b) square kerf structure. In all cases, the incident strain signal was nearly constant at 1000 μstrain. The experimental data in (a) is from Fig. 4 for $\dot{\varepsilon} = 490 \pm 20\ \mathrm{s}^{-1}$

The simulated bar strains in Fig. 7 (b) resulted in the calculated stresses within the square kerf structure shown in the top row of images in Fig. 8. The incident bar-platen is at the left—and transmitted platen-bar at the right—with the annotated arrow indicating the direction of incident wave propagation. The wave is observed to enter the kerf structure until 21 μs and then is reflected at the transmitted platen interface. After 21 μs, the kerf structure is at a nominally uniform and elevated stress level for the remaining duration of the compression pulse. The bottom row of images in Fig. 8 are successive frames of the high-speed video collected over a longer time than the simulation images. While passage of the wave is not experimentally visible (but possibly observable with future coupling of digital image correlation method), a qualitative comparison of the kerf structure deformation in the experimental (150 μs) and simulated (132 us) images show reasonable agreement. Future work will incorporate models breaking of the sample after the ultimate stress is reached which will be necessary to represent the experimental data (e.g., Fig. 8 at 250 μs).

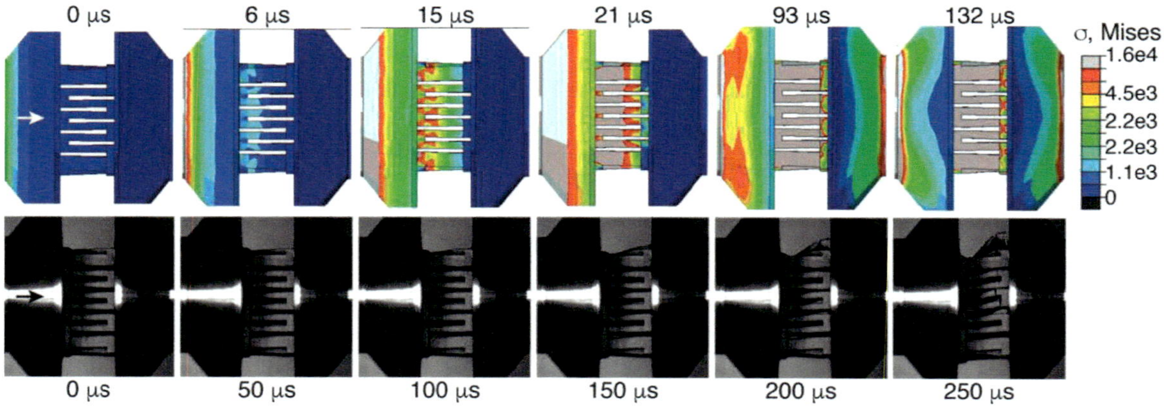

Fig. 8 Top row: simulation results for early times during arrival of the incident compression stress pulse corresponding to the simulated strain data of **Fig. 4** (b). Colorbar corresponds to Mises stresses in psi. Bottom row: a series of successive images captured at 50 MHz corresponding to the experimental strain data of **Fig. 4** (b). Incident bar with platen is at left and the transmitted bar with platen is at right

The deformation behaviors of the square kerf structures are compared in Fig. 9 for the PLA (top row of images) and PF (bottom row of images). Note that the images are provided over a longer time with zero time corresponding to the image before first motion of the incident platen is observed. The kerf structures for both materials fractured although the PF undergoes bending and densification before fracture because it is softer than PLA. PF failure appears to initiate at the long connectors. The stiff PLA does not show connector buckling prior to fracture, rather PLA failure starts at the junction of the connectors and cells. This early investigation presents opportunities to control wave propagation by tuning the materials and geometry of hierarchical kerf structures.

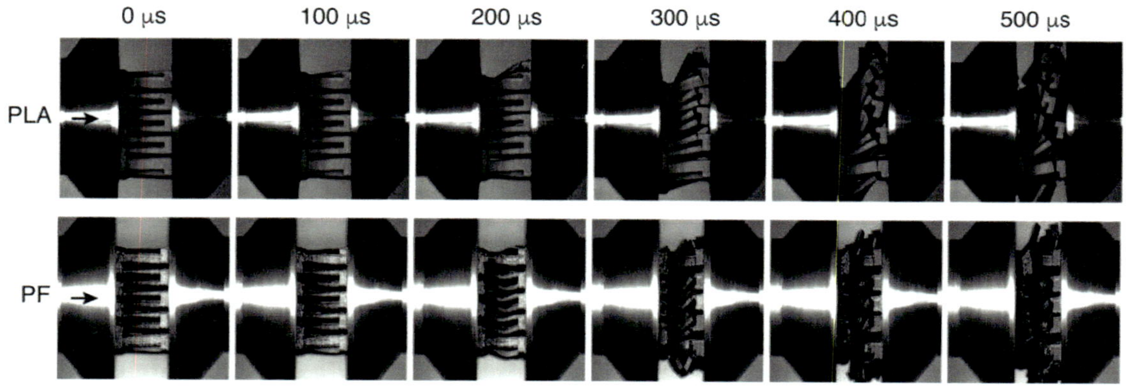

Fig. 9 Top row: a series of successive images during the dynamic compression of the PLA square kerf structure. Bottom row: a series of successive images during the dynamic compression of the PF square kerf structure. Equivalent test conditions with an incident strain of 800 μstrain. Incident bar with platen is at left and the transmitted bar with platen is at right

Conclusions

A collaborative experimental and computational study of energy management in a novel 3D kerf structure is presented. Individual materials of PLA, TPU and a 50:50 mixture by volume of a PF were studied in traditional disk samples for the collection of material stress-strain data and in novel 3D kerf structures. While the PLA exhibited brittle, low strain and high stress to failure and the TPU exhibited compliant, large deformations with recovery, the PF did not exhibit a mean behavior. Rather, the PF was interestingly insensitive to strain rate effects and exhibited behaviors like the more compliant TPU. The PF kerf structure exhibited an earlier onset of severe deformation and fracture as compared to the PLA kerf structure. This work is ongoing with additional investigations exploring the sensitivity of bulk behavior on material and kerf structure design.

Acknowledgments This work was supported by a 2023-2024 MEEN Seed Grant within the J. Mike Walker '66 Department of Mechanical Engineering at Texas A&M University.

References

1. Bianconi F, Filippucci M (2019) Digital wood design: innovative techniques of representation in architectural design. Springer
2. Shahid Z, Bond CG, Johnson MS, et al (2022) Dynamic response of flexible viscoelastic kerf structures of freeform shapes. *Int J Solids Struct* 254–255. https://doi.org/10.1016/j.ijsolstr.2022.111895
3. Shahid Z, Hubbard JE, Kalantar N, Muliana A (2022) An investigation of the dynamic response of architectural kerf structures. *Acta Mech* 233:157–181. https://doi.org/10.1007/s00707-021-03108-z
4. Murray CM, Mao M, Park J, et al (2023) Visco-Elastic Honeycomb Structures with Increased Energy Absorption and Shape Recovery Performance Using Buckling Initiators. *Polymers* (Basel) 15. https://doi.org/10.3390/polym15163350
5. Rudykh S, Ortiz C, Boyce MC (2015) Flexibility and protection by design: Imbricated hybrid microstructures of bio-inspired armor. *Soft Matter* 11:2547–2554. https://doi.org/10.1039/c4sm02907k
6. Nam R, Jakubinek M, Niknam H, et al (2023) 3D printed octet plate-lattices for tunable energy absorption. *Mater Des* 228. https://doi.org/10.1016/j.matdes.2023.111835
7. Kumar A, Verma S, Jeng JY (2020) Supportless lattice structures for energy absorption fabricated by fused deposition modeling. 3D Print *Addit Manuf* 7:85–96. https://doi.org/10.1089/3dp.2019.0089
8. Tsouknidas A, Pantazopoulos M, Katsoulis I, et al (2016) Impact absorption capacity of 3D-printed components fabricated by fused deposition modelling. *Mater Des* 102:41–44
9. Gray III GT (2002) High-Strain-Rate Testing of Materials. Characterization of Materials
10. Chen WW, Song B (2010) Split Hopkinson (Kolsky) bar: design, testing and applications. Springer Science & Business Media
11. Darnal A, Shahid Z, Deshpande H, Kim J, and Muliana A, (2023) Tuning Mechanical Properties of 3D Printed Composites with PLA:TPU Programmable Filaments. *Composite Structures*, 318,117075.

Chapter 8
Dynamic four-point bend fracture with a modified Split Hopkinson (Kolsky) Bar

Aishvarya Joshi, Cody Kirk, Tyler Dillard, Matthew Kappes, Zherui Martinez-Guo

Abstract High-rate deformation and fracture behavior of aerospace-grade alloys, such as Ti-6Al-4V and Al 2024, are important for understanding their performance under extreme loading conditions. The objective of our study is to design an experimental setup and investigate the dynamic fracture behavior of these alloys using a modified Split Hopkinson Pressure Bar (SHPB) setup integrated with a four-point bend fixture to evaluate Single Edge Notched Bend (SENB) specimens with different triaxiality are tested to analyze their respective fracture strain. The experiments are conducted for high-strain-rate conditions, with Digital Image Correlation (DIC) used to measure local strain fields and ABAQUS simulations to determine stress triaxiality. Pulse shaping techniques are implemented to ensure dynamic equilibrium and minimize inertial effects. The results would demonstrate a dependence of fracture behavior on notch geometry and strain rate, providing valuable data for developing high-fidelity material models that predict failure across various geometries.

Keywords SHPB · Triaxiality · Fracture · Bending · Material model

Introduction

Titanium and aluminium alloys are widely used in the aerospace industry due to their exceptional properties that meet the stringent requirements of aircraft and spacecraft. Among the most widely utilized alloys, Ti-6Al-4V and Al 2024 are particularly favoured for their high strength, durability, and resistance to extreme forces and structural damage with low weights, making them ideal for aerospace applications, especially for turbine blades which endure extreme environmental conditions under high speeds. Ti-6Al-4V is an $\alpha+\beta$ phase alloy, which is its most stable form of Ti-6Al-4V alloy, provides high strength when heat-treated [3]. Similarly, Al 2024-T351 is used for its high strength-to-weight ratio [4]. Despite Ti-6Al-4V and Al 2024's wide usage in the aerospace applications, there is a lack of understanding of the fracture and crack initiation for local features. The failure progression in Ti-6Al-4V and Al 2024 under high strain-rate loading begins with elastic deformation, where experimental force-displacement data and strain measurements show strong agreement with numerical simulations. As the load increases, the material transitions into plastic yielding, where localized plastic strain accumulation at the notch tip begins to develop. Following plastic deformation, crack initiation occurs at regions of maximum stress concentration, which shows variation in from standard behavior in simulations due to strain rate sensitivity and microstructural heterogeneity. This emphasizes the need for refined failure models to better capture the transition to fracture. Hence, it is important to evaluate the changes in their material properties and performance at such conditions.

Lesuer et al. evaluated the ability of the Johnson-Cook model to represent the deformation and failure response of these materials under conditions relevant to simulations of engine containment and the influence of uncontained engine debris on aircraft structures [5]. Despite challenges such as time-consuming calibration, failure criteria simulations are gaining traction in the aerospace, automotive, and manufacturing industries [3] [4]. These failure criteria enable impact damage simulation and residual integrity assessment, aiding in aerospace crashworthiness analysis.

Aishvarya Joshi (✉) · Cody Kirk · Tyler Dillard · Zherui Martinez-Guo
School of Aeronautics & Astronautics Engineering, Purdue University, West Lafayette, IN, USA
e-mail: joshi270@purdue.edu; kirk24@purdue.edu; tdillard@purdue.edu; zmg@purdue.edu

Zherui Martinez-Guo
School of Materials Science Engineering, Purdue University, West Lafayette, IN, USA
e-mail: zmg@purdue.edu

Matthew Kappes
Rolls-Royce Corporation, 450 South Meridian Street, Indianapolis, IN, USA
e-mail: Matthew.Kappes@Rolls-Royce.com

Paul Allison et al. (eds.), *Dynamic Behavior of Materials, Vol. 1*, Conference Proceedings
of the Society for Experimental Mechanics Series, https://doi.org/10.13052/9788743808275-8

One of the failure criteria that researchers have explored is the Bao–Wierzbicki framework which further enhances the predictive capabilities of failure models, and it incorporates stress triaxiality as a key parameter. Fracture calibration of Ti-6Al-4V and Al 2024 using this framework to characterize ductile failure criteria represents a state-of-the-art approach based on stress triaxiality. Since stress triaxiality depends on the geometry of the specimen, this method provides an effective way to correlate geometry with material properties and predict failure [6]. The development and refinement of advanced material models for Al 2024 T3 and Ti-6Al-4V have also been done by DOT/FAA/TC-13/25 P1, P2, P3 [15][16][17] and DOT/FAA/TC-15/23 [18] respectively where material model parameters for input in LS-DYNA for simulation are in parallel with mechanical experiments under various strain rates. Damage due to tensile and compression, high energy - dynamic loading on both materials is analyzed and reported to create a material model for different loading conditions. Metal plasticity and fracture modeling methodology with the help of experimental data is also described in this research [19][20][21]y. Evaluations using hat-shaped specimens for the stress triaxiality and fracture strain have been conducted by Wang et al. [7]. However, crack initiation at high triaxialities under high strain rates in Single Edge Notched Bend (SENB) specimens is a field that provides an opportunity for further research.

Experimental techniques such as the Split Hopkinson Pressure Bar (SHPB) provide a controlled method for investigating these fracture phenomena under high strain rates. A Split Hopkinson Pressure Bar, also called as the Kolsky bar, is an experimental setup used to conduct experiments under high rate loading under laboratory-controlled conditions to obtain dynamic response of materials. In this setup, two elastic rods were placed on either side, with one struck by the striker under pressure from the gas gun. As shown in the schematic Figure 2, the rod between the striker and specimen is the incident bar, while the rod on the other side is the transmission bar. When the incident bar is struck, a compressive stress wave propagates toward the specimen, moving the bar material as it passes. Upon reaching the interface, part of the wave reflects into the incident bar, while the rest transmits through the specimen into the transmission bar. Laboratory instrumentation records these waves—both the incident wave and its reflection, as well as the wave in the transmission bar—allowing for a consistent and quantitative impact event. By varying the impact velocity and specimen size, different strain rates can be achieved. Further analysis of the recorded waves provides insights into loading conditions and specimen deformation states [2].

To dynamically load and evaluate the Single Edged Notch Bend Specimen or SEN(B) specimen under high strain rates, an SHPB setup was modified incorporating four-point bend fixtures as shown in Figure 2. This kind of setup has been illustrated in Weerasooriya et al. and Henschel et al. [1] [8]. This study focuses on evaluating Ti-6Al-4V and Al 2024 at high strain rates for different triaxiality values. Fracture strain is determined using Digital Image Correlation (DIC), while triaxiality factors are obtained through ABAQUS simulations. These results are correlated to develop a material model capable of predicting fracture behavior at high strain rates across various geometries.

Background

Specimen Configuration

The specimen geometry was designed in accordance with ASTM E399-23 (*Standard Test Method for Linear-Elastic Plane-Strain Fracture Toughness of Metallic Materials*) [9], specifically Annex A3.2, which outlines special requirements for testing Single Edge Notched Bend (SENB) specimens. The specimen dimensions were scaled down in accordance with the Kolsky bar dimensions available in our facility. Additionally, ASTM E23-34 (*Standard Test Methods for Notched Bar Impact Testing of Metallic Materials*) [10], which provides specifications for impact testing of U – notch specimens, was referenced for the design of U notches with the SEN(B) configuration with variation in triaxiality.

Table 1 presents the calculated parameters derived from these standards. Simulation results indicated no significant variation in the stress triaxiality factor for notches with equal radii, highlighting notch radius as the primary factor influencing triaxiality variation across specimens. Accordingly, five different notch radii, ranging from 0.15 mm to 1.5 mm, were selected for testing. Figure 1 illustrates the specimen dimensions.

The spacing of the support pins and load pins were according to ASTM C1161-18 [11]. The fixtures are made of A2 Steel heat treated according to ASTM A681-08 [12]y. The hardness of the fixtures measured after the heat treatment was 62 HRC. Hertzian Analysis was done to predict the elastic indentation stress limit for the hardened fixtures which was calculated to be 2.41GPa [13]. To minimize the wave disturbances caused by the introduction of the fixtures, the wave impedance of the fixtures should match with those of the bars, $(\rho cA)_{fixtures} = (\rho cA)_{bar}$ to minimize the wave disturbances caused by the introduction of the fixtures [2]. Since the density and wave speed in Kolsky bars and fixtures is the same, the cross-section area of the fixtures was matched to that of the Kolsky bars. The fixtures were attached to the Kolsky bar using support collars made of Aluminum 6061-T6.

Table 1 Dimensions calculation based on ASTM E399-23 standards

Parameter	Value (mm)	Criterion
Width W	9	Size of Kolsky Bar
Thickness B	4.5	B $= W/2$
Length L	37.8	L $= 4.2$W
Notch length a	4.1	$0.45 \leq a/W \leq 0.55$
Notch Radii R	0.15, 0.3, 0.65, 1, 1.5	Stress triaxiality variation

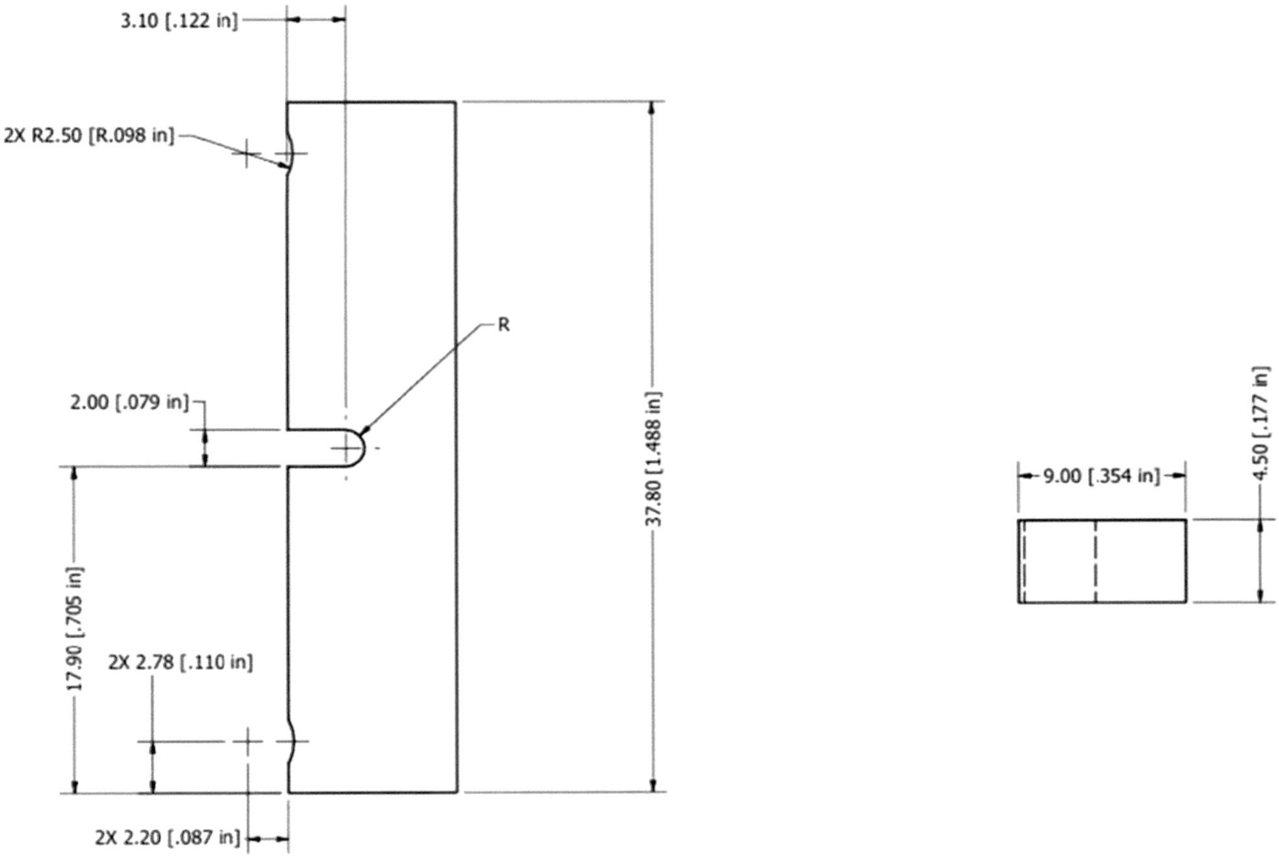

Fig. 1 Specimen Configuration with R varying as 0.15mm, 0.3mm, 0.65mm, 1mm and 1.5mm

High-Rate Testing Setup

The Kolsky Bar in our experimental setup is made of C300 Steel. The wave speed in the bar is calculated as $C_{bar} = \sqrt{\frac{E}{\rho}}$ where E $=$ E_{bar}(Elastic modulus of C300 Steel) and $\rho = \rho_{bar}$(Density of C300 Steel). Ambient temperature is maintained for the experiments.

The specimen undergoing fracture needs to be in dynamic equilibrium and needs to fracture at a constant strain rate. It is crucial to initiate the crack in the specimen at the first pulse to get accurate measurements of local fracture strain at the constant strain rate applied. To have a gradual ramp in loading and give enough time for the specimen to come in equilibrium, a copper pulse shaper of thickness 4.7mm and the diameter $\frac{1}{4}$ inches was used. The velocity of incident bar was 9.83 m/s. Figure 4 shows the ramp loading and the duration of incident loading in two different pulse shapers where initially a thinner copper pulse shaper was used (thickness = 1.7mm). However, the loading duration was too short for that. Eventually, increasing the thickness of the pulse shaper to 4.7mm gave us the incident loading we wanted,

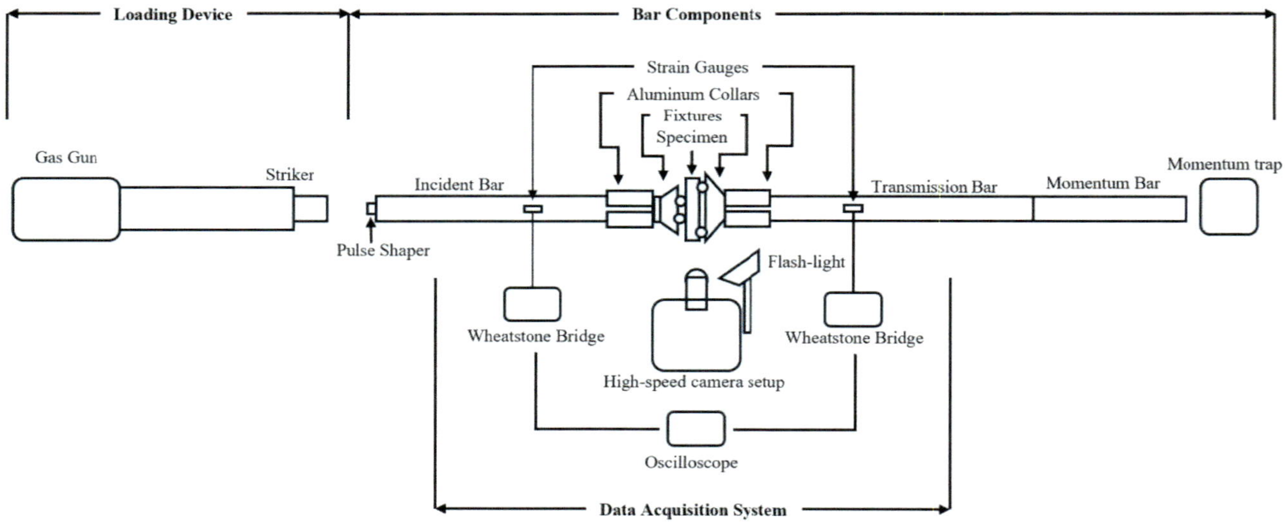

Fig. 2 Schematic of the experimental setup

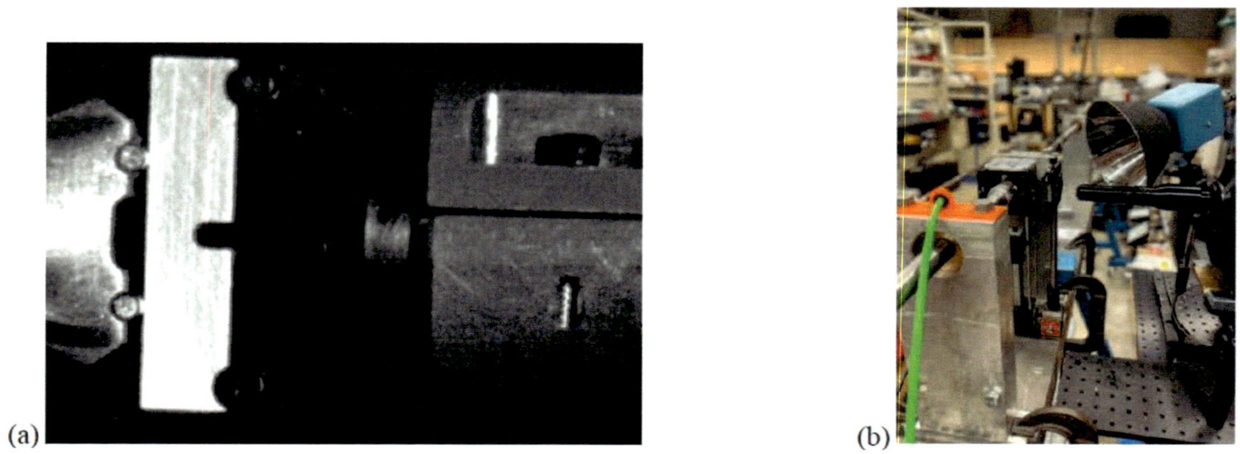

Fig. 3 (a) 4-point bend setup (b) Kolsky bar with Shimadzu Camera and Flash Lamp setup

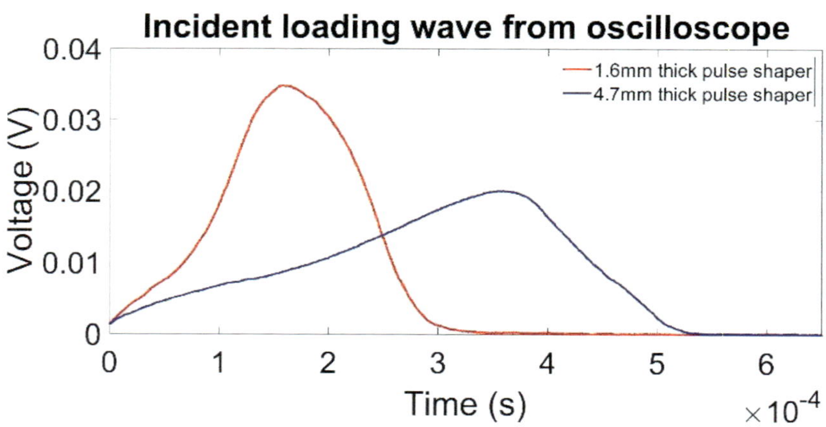

Fig. 4 Plot depicting increase in loading duration due to change in thickness of pulse shaper

Analysis

Resonant frequency analysis

To determine the upper limit in loading rate under which equilibrium is ensured, it is necessary to determine the first resonant frequency of the specimen within the specific boundaries of the experimental setup. If the highest frequency component of the loading pulse is much higher or lower than the first resonant frequency, the specimen is approximately in its equilibrium state. To estimate the first resonant frequency of the specimen in the modified SHPB setup, the experimental configuration is modeled as a multi-body dynamic system, with reference to Weerasooriya et al [1]. In this simplified model, the specimen is represented by a single-degree mass-spring system with mass m and elastic constants K1 and K2. A1 and A2 are cross sectional areas, E1 and E2 are Young's moduli, and $\rho 1$ and $\rho 2$ are material densities of the incident bar and the material respectively. The forces F1 and F2 are loads acting on both sides of the specimen. A dynamics analysis with details given in [1], reveals that the transfer function between F1and F2 has the following form in equation E1:

$$H(\omega_n) = \frac{F_2}{F_1} = \frac{1}{1 - \frac{m\omega_n^2}{K_1} + i\frac{m\omega_n^2}{K_2 E_2 A_2}} \tag{E1}$$

As shown in Figure 5 (a) and 5(b), the first resonant frequency is 93.2 kHz. The base frequency of the loading pulse is estimated by performing FFT (Fast Fourier Transform) on the loading signal and estimated it to be around 2000Hz as shown in Figure 6 (a) and 6(b), which is below the resonant frequency. The transfer function H thus has a value close to unity as can be observed in Figure 5 (a) and 5 (b) at all frequency values before the resonant frequency.

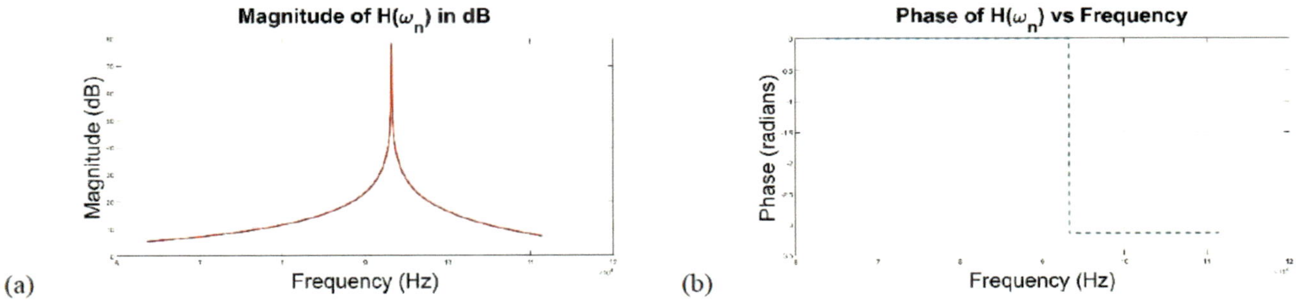

Fig. 5 (a) and (b) are plots of H vs Frequency depicting resonance frequency at 93.2kHz.

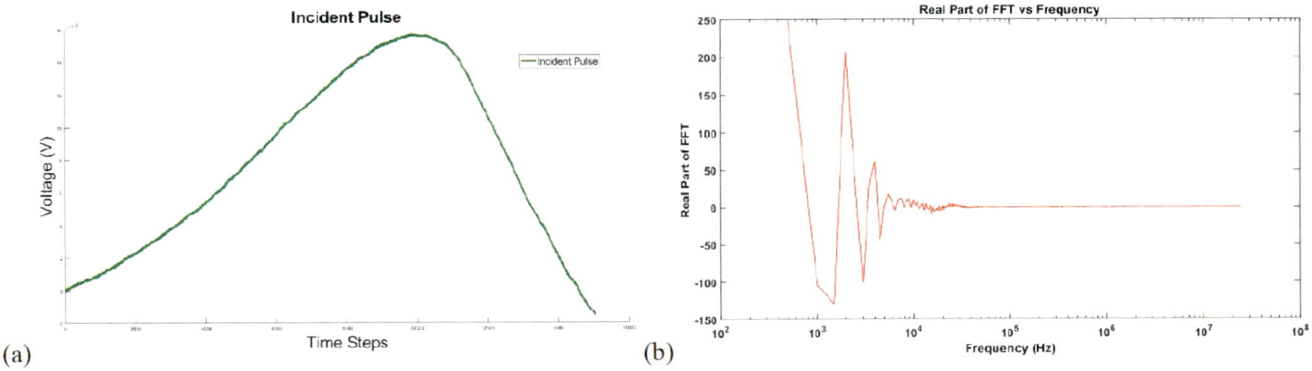

Fig. 6 (a) Incident load pulse (b) real part of FFT of incident signal to determine base frequency.

Measurements

The Kolsky bar setup consists of loading device which contains the gas gun, striker bar, incident bar, transmission bar, and momentum bar as shown in Figure 2. Strain gauges mounted on both bars measure the incident, reflected, and transmitted strain waves with the help of Wheatstone bridges and oscilloscope data [2]. To verify the viability of this setup, initial studies were performed on an Al 2024 specimen with a notch radius of 1.5mm. Figure 6 shows the overlay incident, reflected and transmitted signal.

The plot in Figure 8 determines the balance of stresses in incident and the transmitted bar. We can see a slight mismatch due to energy dissipation from the fixtures of four-point bend setup attached to the Kolsky Bar.

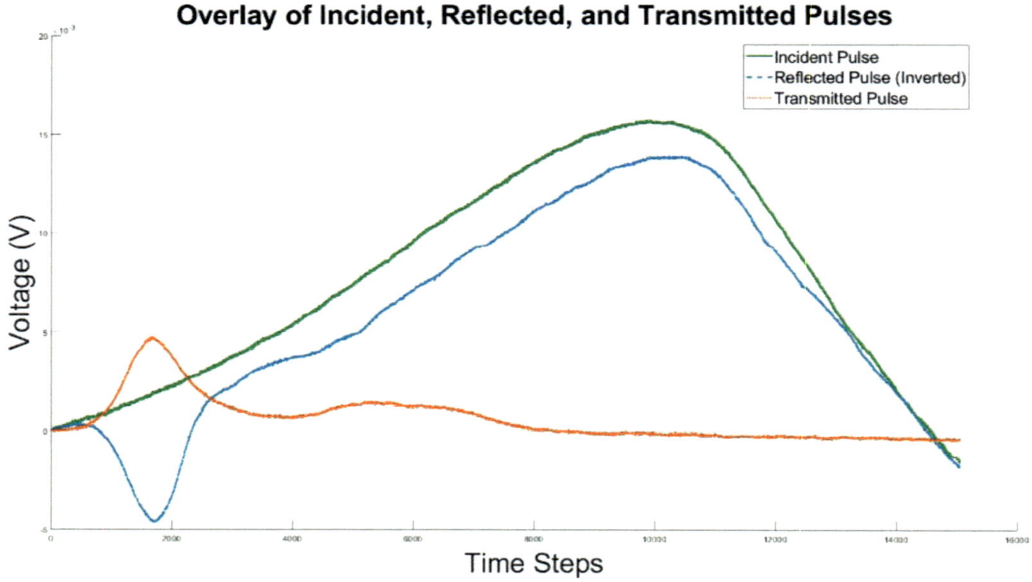

Fig. 7 The recorded incident, reflected and transmitted pulse data from the oscilloscope

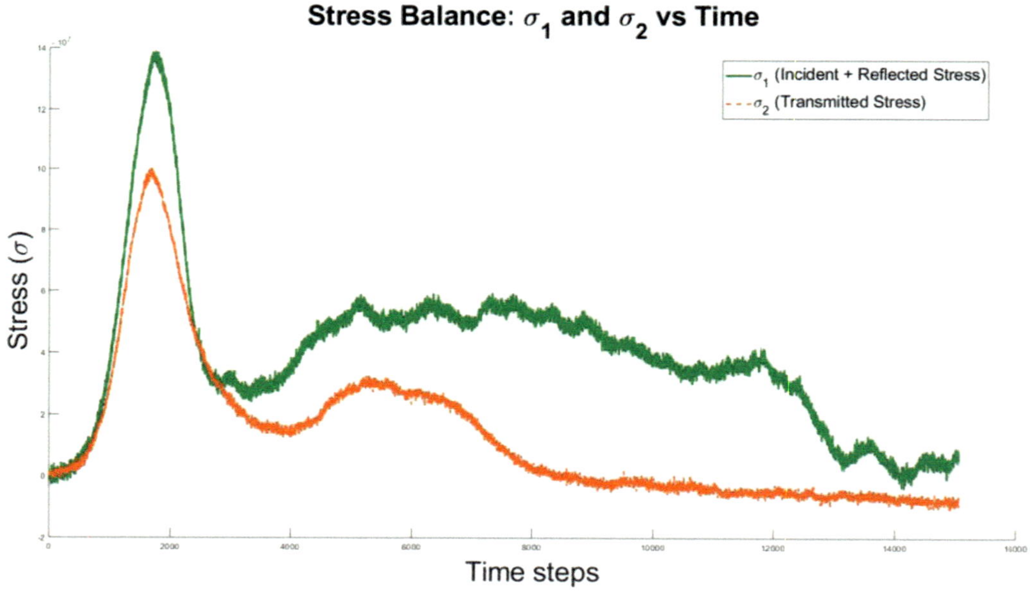

Fig. 8 Stress Balance to determine equilibrium

A high-speed camera is used to monitor the cracking of the specimen, the captured images are then used to perform DIC (Digital Image Correlation) to calculate local deformation. High speed images were captured using a Shimadzu HPV-X2 camera capable of up to 10 million frames per second (10 MHz framerate). The camera is orthogonally positioned to the bar axis. The HPV-X2 has a frame resolution of 400 $\times$ 250 pixels and records a maximum of 128 frames regardless of framerate.

A single Cordin Model 659 capacitive discharge flash lamp which was mounted beside the camera was used for consistent illumination. Delay timings are adjusted within the Shimadzu software to account for time taken by elastic wave propagated from strain gauge to the sample. Recorded camera frames are output from the HPV-X2 into the oscilloscope to correlate the captured image sequence frames to the corresponding load signals for post-impact analysis. The images from Shimadzu can be seen in Figure 9 at different loading conditions. The peak load calculated was 95483.21 N.

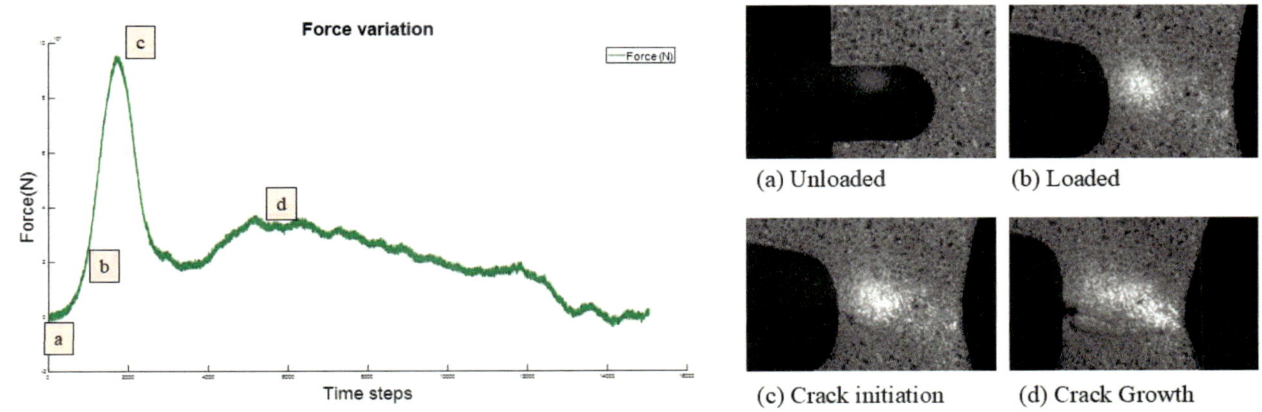

Fig. 9 Force and loading conditions

Triaxiality is an important parameter to determine material behavior under fracture. It is the ratio of hydrostatic stress to Von-Mises stress. The value determines the resistance for crack growth. Hence, higher stress triaxiality relates to brittle failure and cleavage and lower stress triaxiality values relate to micro-void coalescence and shear lip formation [14]. Quasi-static analysis was carried out for all geometries to evaluate the quasi-static variation of triaxiality factors at the notch tips respectively. The material model is taken as Johson-Cook Plasticity model with values from Leuser et al [5]. These values can be seen in the trend shown in Figure 10.

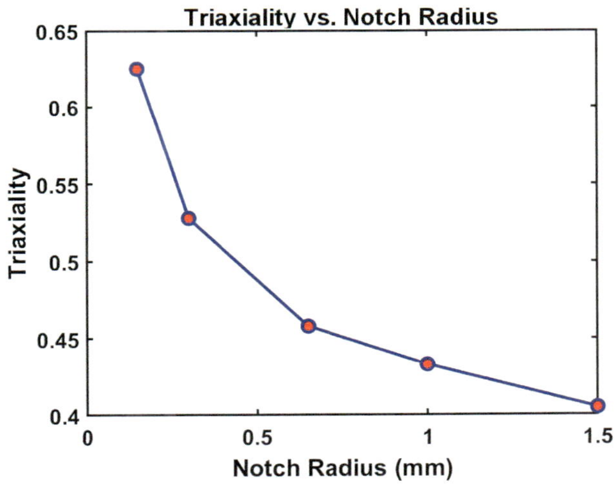

Fig. 10 Variation of stress triaxiality with notch radius from ABAQUS simulations.

Future Work

Specimens with the mentioned notch geometries for Al 2024 and Ti-6Al-4V will be conducted with the given experimental setup. The strain gauge data from the Kolsky Bar will be used to understand the loading conditions, calculate strain rate and the velocity of impact. DIC will be employed to evaluate and post process experimental data from high-speed camera to obtain fracture strain values. Simulations will be carried out in ABAQUS to obtain the stress triaxiality values. Correlation of fracture strain and triaxiality values will be done to establish a material model for elastic, plastic and crack initiation regimes for improved fracture predictions.

Conclusion

This study investigates Ti-6Al-4V and Al 2024 in a detailed experimental setup for the investigation of high-strain-rate fracture behavior using a modified SHPB setup with four-point bend fixtures. While the Johnson-Cook model has been used to simulate their deformation and failure, crack initiation at high triaxialities under high strain rates in Single Edge Notched Bend (SENB) specimens is conducted to establish a relationship between stress triaxiality and fracture strain at high strain rates. The notch sizes are varied to vary the stress triaxiality and the specimens are fractured in a single pulse of incident force to calculate fracture strain using DIC. To ensure dynamic equilibrium in the specimen, pulse shaping techniques are employed. As a part of future work, experiments will be conducted on Al 2024 and Ti-6Al-4V for different notch radii to identify fracture strain with the help of experiments and stress triaxiality with simulations and establish a relationship between them to create material models at high strain rates.

References

1. Weerasooriya, Tusit, Paul Moy, Daniel Casem, Ming Cheng, and Weinong Chen. "A four-point bend technique to determine dynamic fracture toughness of ceramics." *Journal of the American Ceramic Society* 89, no. 3 (2006): 990-995.
2. Chen, Weinong W., and Bo Song. *Split Hopkinson (Kolsky) bar: design, testing and applications*. Springer Science & Business Media, 2010.
3. Williams, James C., and Rodney R. Boyer. "Opportunities and issues in the application of titanium alloys for aerospace components." *Metals* 10, no. 6 (2020): 705.
4. Li, Shuang–Shuang, Xin Yue, Qing–Yuan Li, He–Li Peng, Bai–Xin Dong, Tian–Shu Liu, Hong–Yu Yang et al. "Development and applications of aluminum alloys for aerospace industry." *Journal of Materials Research and Technology* (2023).
5. Lesuer, D. R. "Experimental investigations of material models for Ti-6Al-4V titanium and 2024-T3 aluminum. US Department of Transportation/Federal Aviation Administration." *DTFA03–97Z-90007* (2000).
6. Wierzbicki, Tomasz, Yingbin Bao, Young-Woong Lee, and Yuanli Bai. "Calibration and evaluation of seven fracture models." *International Journal of Mechanical Sciences* 47, no. 4-5 (2005): 719-743.
7. Wang, Bing, Xinran Xiao, Viktor P. Astakhov, and Zhanqiang Liu. "The effects of stress triaxiality and strain rate on the fracture strain of Ti6Al4V." *Engineering Fracture Mechanics* 219 (2019): 106627.
8. Henschel, Sebastian, and Lutz Krüger. "Crack initiation at high loading rates applying the four-point bending split Hopkinson pressure bar technique." *international Journal of Fracture* 201 (2016): 235-248.
9. ASTM International Committee E08 on Fatigue and Fracture. Subcommittee E08. 07 on Fracture Mechanics. *Standard Test Method for Linear-elastic Plane-strain Fracture Toughness K [Ic] of Metallic Materials*. ASTM international, 2023.
10. ASTM Committee E28 on Mechanical Testing and Subcommittee E28.07 on Impact Testing. *Standard Test Methods for Notched Bar Impact Testing of Metallic Materials*. ASTM international, 2024.
11. ASTM Committee C28 on Advanced Ceramics and Subcommittee C28.01 on Mechanical Properties and Performance. *Standard Test Method for Flexural Strength of Advanced Ceramics at Ambient Temperature*. ASTM international, 2018.
12. ASTM Committee A01 on Steel, Stainless Steel and Related Alloys and Subcommittee A01.29 on Tool Steels. *Standard Speci?cation for Tool Steels Alloy*. ASTM international, 2008.
13. Fischer-Cripps, Anthony C. *Introduction to contact mechanics*. Vol. 101. New York: Springer, 2007.
14. Anderson, Ted L., and Ted L. Anderson. *Fracture mechanics: fundamentals and applications*. CRC press, 2005.
15. Emmerling, W., D. Altobelli, K. Carney, and M. Pereira. "Development of a new metal material model in LS-DYNA Part 1: FAA, NASA, and industry collaboration background [DOT/FAA/TC-13/25]." *Federal Aviation Administration* (2014).
16. Buyuk M. *Development of A tabulated thermo-viscoplastic material model with regularized failure for dynamic ductile failure prediction of structures under impact loading*. [Order No. 3597748]. The George Washington University; 2013. *DOT/FAA/TC-13/25 Part 2*
17. Seidt, Jeremy Daniel. "Plastic deformation and ductile fracture of 2024-T351 aluminum under various loading conditions." PhD diss., The Ohio State University, 2010. *DOT/FAA/TC-13/25 Part 3*
18. Haight, S., L. Wang, and P. Du Bois. *Development of a titanium alloy Ti-6Al-4V material model used in LS-DYNA DOT*. FAA/TC-15/23, 2016.
19. Haight, Sean Harrison. "An anisotropic and asymmetric Material Model for Simulation of Metals under dynamic Loading." PhD diss., George Mason University, 2016. *DOT/FAA/TC-TT16/44*
20. Hammer, Jeremiah Thomas. "Plastic deformation and ductile fracture of Ti-6Al-4V under various loading conditions." Master's thesis, The Ohio State University, 2012. *DOT/FAA/TC-TT14/2*
21. Lesuer, Donald R. "Experimental investigations of material models for Ti-6Al-4V titanium and 2024-T3 aluminum." *DOT/FAA/R-00/25* (2000).

Chapter 9
Development of a Novel Method for Simultaneous Measurements of Multiple Constant Elevated Strain Rates

Derryk Daignault, Tyler Robertson, and Nathan Spulak

Abstract A key challenge in dynamic testing is the fact that the material specimens exhibit non-uniform deformation and changing geometries throughout the experiment. This creates variations in the instantaneous true strain rate of the material during the experiment, making it difficult to accurately characterize the material rate sensitivity. To address this, novel high rate split-Hopkinson bar experiments are designed using finite element analysis with Ansys LS-DYNA, taking advantage of strain localization in order to simultaneously induce multiple constant strain rates on a single experimental specimen. Use of these novel test designs and analysis methods will enable the stress vs. strain behavior at multiple truly constant strain rates to be obtained from a single experiment, allowing for both more efficient and more accurate characterization of material rate sensitivity.

Keywords Strain rate · Strain localization · Split-Hopkinson bar · Digital image correlation · Finite element analysis

INTRODUCTION

It is necessary to characterize material deformation and fracture behavior at high strain rates for a variety of applications, including vehicle crashworthiness and high velocity impacts, in order to accurately design structures for use in such scenarios. A key experimental technique for determining the material response at elevated strain rates is the split-Hopkinson bar (SHB) technique, which is capable of loading specimens at elevated strain rates on the order of 10^3 s^{-1} under dynamic force equilibrium and with uniform engineering strain rates [1]. Yet SHB experiments are typically not capable of achieving constant true strain rates on the material specimen, as shown in Figure 1 [2].

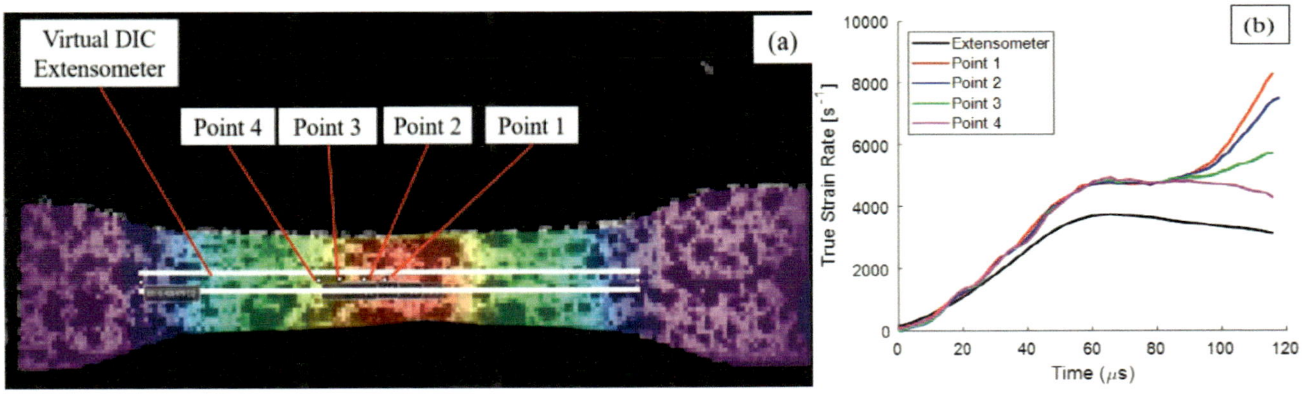

Fig. 1 Tension split Hopkinson bar experiment on additively manufactured 316L stainless steel. The true strain rate does not remain constant with respect to time and varies significantly across the specimen once necking occurs

Derryk Daignault · Tyler Robertson · Nathan Spulak
Department of Mechanical and Aerospace Engineering, University of Alabama in Huntsville, Huntsville, AL 35899
e-mail: dd0132@uah.edu; tr0085@uah.edu; nathan.spulak@uah.edu

Paul Allison et al. (eds.), *Dynamic Behavior of Materials, Vol. 1*, Conference Proceedings
of the Society for Experimental Mechanics Series, https://doi.org/10.13052/9788743808275-9

This is due in part to the changing specimen gauge length for either tension and compression tests, which will result in a variation of the true strain rate as the test progresses. However for tests where strain localization occurs, such as necking in tension tests, the localized strain rate drastically increases, resulting in strain rates that are not constant with respect to time or spatially across the specimen. Typical SHB analysis methods utilizing solely the average strain rate may not fully capture the material's deformation response and rate sensitivity due to the multiple strain rates that occur during such tests. Therefore a novel approach taking advantage of this strain localization phenomenon has been pursued, to generate multiple different localized true strain rates across the specimen that are each constant with respect to time.

EXPERIMENTAL DESIGN

Multiple tension test specimen designs are evaluated using the finite element analysis (FEA) software Ansys LS-DYNA. The specimens are desired to have immediate strain localization from the onset of the test, and a constant localization rate such that the true strain rate remains constant at discrete locations along the specimen. Example specimen geometries are shown in Figure 2, with either a smooth gauge (SG) variation or a sharp V-notched (VN) gauge variation.

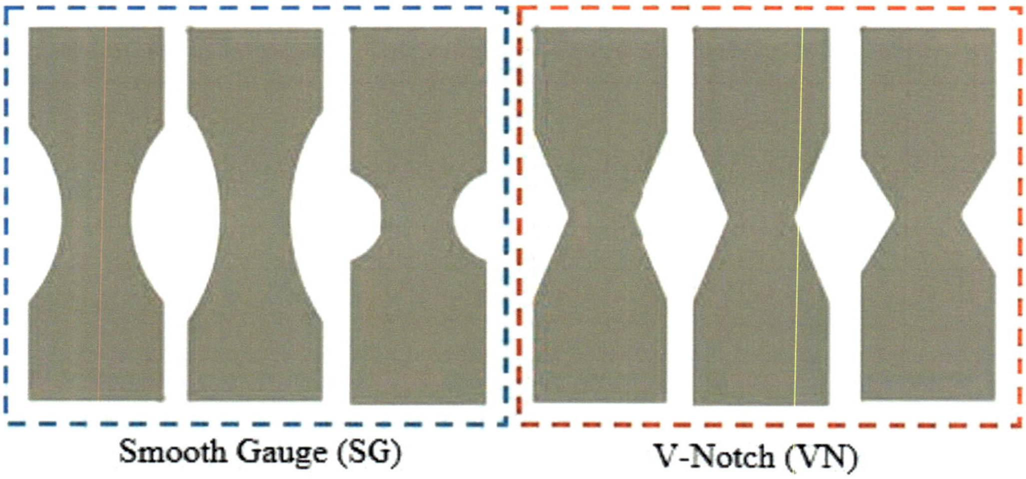

Fig. 2 Examples of specimen geometries with varying cross-sectional geometries, to induce immediate strain localization and constant true strain rates at discrete locations along specimen

Each specimen design is simulated, using the strain rate sensitive MAT-224 material model in LS-DYNA developed through testing of titanium 4Al-6V at various strain rates [3]. Simulations are performed with a mesh size of 0.2 mm, and two sets of boundary conditions are utilized. One boundary condition set utilizes a fully fixed boundary condition on one end of the specimen coupled with a constant velocity displacement on the other end to achieve tensile loading, representative of the experimental conditions for tests on hydraulic load frames. The other boundary condition set utilizes both ends of the specimen moving to represent the conditions observed in an SHB experiment, with the velocity profiles assigned to either ends of the specimen taken from representative SHB experimental data [2][4].

RESULTS

The simulated strain rate with respect to time at various points along a representative SG and a VN specimen is shown in Figure 3. Each line represents the local true strain rate at a single element along the centerline of the specimen gauge length. As can be seen, the simulations predict a range of strain rates across their length, with the peak strain rates occurring towards the center of each specimen where the gauge width is narrowest.

Neither specimen design achieves truly constant true strain rates at all locations. However there are specific locations where the simulated true strain rate does remain at a constant level as the test progresses, typically at a region just outside of the specimen center. The VN specimen geometry demonstrates a similar behavior as the standard dogbone tension specimen in Figure 1, with relatively constant strain rates initially followed by a rapid increase in the localized strain rate after the

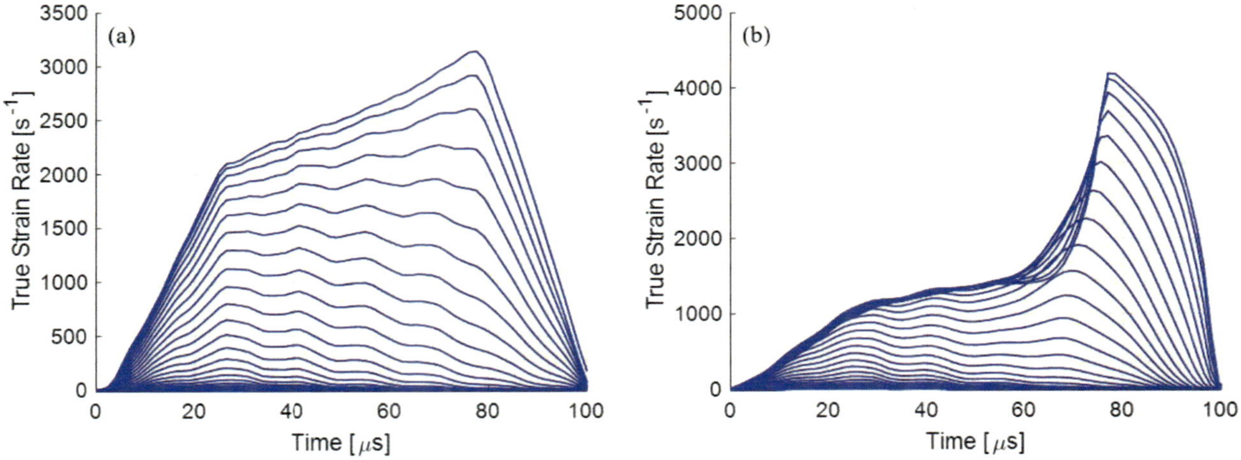

Fig. 3 Simulated strain rate at various discrete locations along specimen gauge length for (a) an SG type specimen design and (b) a VN type specimen design

onset of strain localization in the center. By contrast the SG specimen geometry shows less variation of the strain rate with respect to time as the test progresses, indicating a more consistent rate of strain localization during the test.

CONCLUSION

The SG and VN specimen designs show promising variation in the strain rate and the ability to achieve a constant true strain rate with respect to time at discrete locations on the material specimen. Future work involves performing experiments using these geometries at a variety of strain rates, and using digital image correlation (DIC) to determine the localized strains and strain rates across the specimen. The corresponding full-field stresses can be analyzed either through measurements of the changes to the local cross-section via DIC, or through advanced analytical methods such as the virtual fields method (VFM). These novel tests demonstrates a promising capability to measure high strain rate material deformation in a more accurate and more efficient manner.

Acknowledgments This work was performed with the support of the University of Alabama in Huntsville's (UAH) Research or Creative Experiences for Undergraduates (RCEU) program.

References

1. Kolsky, H. (1949). An investigation of the mechanical properties of materials at very high rates of loading. *Proceedings of the physical society. Section B*, *62*(11), 676.
2. Marler, T. E. (2024). *Quasi-static and dynamic tension testing of as-built and heat-treated additively manufactured 316L stainless steel*. The University of Alabama in Huntsville.
3. Haight, S., Wang, L., Du Bois, P., Carney, K., & Kan, C. D. (2016). Development of a titanium alloy Ti-6Al-4V material model used in LS-DYNA. *George Mason University, The George Washington University, NASA Glenn Research Center, Washington, DC, Report.*
4. Seidt, J. D. (2010). *Plastic deformation and ductile fracture of 2024-T351 aluminum under various loading conditions* (Doctoral dissertation, The Ohio State University).

Chapter 10
Effect of Thermal and Mechanical Fatigue Loading on the Performance of Energetic Materials

Addis Kidane, Suraj Ravindran, Georgios Gavris, and Waiching Sun

Abstract Energetic materials widely used in defense and civilian applications are complex material systems incorporating binders, crystals, and other additives. Their failure mechanisms are complex and depend on many parameters, such as initial porosity, temperature, and loading rate. During service and transportation, the temperature and mechanical loading fluctuation can initiate and accumulate damage and degreed the material's mechanical strength. The effect of fatigue damage on the performance of energetic material needs to be better understood. This required understanding how the damage is accumulated in this material under different temperatures and loading conditions.

We propose a digital image correlation-based experiment to understand the mechanical and thermal fatigue damage accumulation in polymer-bonded energetic materials. This experimentation will be combined with first-principles physics-based multi-scale modeling to establish load-structure-damage-performance linkages for pressed and plastic-bonded energetic materials.

We used polymer-bonded sugar as a surrogate sample and subjected it to controlled fatigue and thermal loading. The microstructures of the sample before and after loading at different loading and thermal cycles are documented and used to study the damage accumulation. The virgin and damaged specimens are further exposed to high-strain loading, and their failure mechanism as a function of damage is also investigated.

Keywords Energetic materials · Thermal · Mechanical · Fatigue

Introduction

Energetic materials (EMs) such as explosives and propellants are designed for rapid energy release. However, thermal and mechanical fatigue can significantly impact their performance, compromising their stability and functionality under operational conditions. Thermal fatigue occurs when EMs are exposed to cyclic temperature changes, leading to the expansion and contraction of the material. This thermal cycling creates internal stresses, resulting in cracks or other forms of structural degradation. Over time, this can cause a decrease in the material's stability and energy release efficiency. Studies indicate that temperature fluctuations can accelerate the degradation of energetic materials, mainly through changes in molecular structure and composition [1, 2]

Mechanical fatigue involves repeated stress or strain on energetic materials, often through cyclic loading or impact. This repetitive mechanical stress can cause physical damage such as cracks, deformation, or delamination. As the material undergoes fatigue, its mechanical properties, including strength and elasticity, can deteriorate, leading to an increased risk of failure. Mechanical stress catalyzes the formation of defects, further compromising the integrity of the material [3-6].

The degradation process can accelerate when thermal and mechanical fatigue are applied simultaneously. The combination of thermal cycling and mechanical loading can lead to a faster breakdown of the material's structure, as these stresses interact and compound one another. The synergistic effects of these factors result in more rapid failure and greater performance loss, including diminished energy release and increased instability [7, 8]

Addis Kidane · Georgios Gavris · Waiching Sun
Department Civil Engineering and Engineering Mechanics, Columbia University, New York, YN 10027
e-mail: addis@civil.columbia.edu; gb2793@columbia.edu; wsun@columbia.edu

Suraj Ravindran
Aerospace Engineering and Mechanics, University of Minnesota, Minneapolis, MN 55455
e-mail: sravi@umn.edu

Paul Allison et al. (eds.), *Dynamic Behavior of Materials, Vol. 1*, Conference Proceedings
of the Society for Experimental Mechanics Series, https://doi.org/10.13052/9788743808275-10

The combined thermal and mechanical fatigue significantly reduces energetic materials' operational lifespan and safety. As fatigue sets in, the material's reliability decreases, leading to potential safety hazards in practical applications such as propulsion systems and detonators. Improving fatigue resistance through new formulations, protective coatings, and enhanced manufacturing techniques can improve the safety of these materials.

Numerical and experimental works have been proposed to understand the failure mechanisms in energetic materials. Researchers, including our group, have developed experiment methods to measure the local deformation mechanism, the compaction wave propagation, and the effect of strain rate on the mechanical properties of energetic materials [11-16].

These studies have contributed to understanding failure in energetic materials, mainly polymer-bonded explosives (PBXs). However, these studies focused on dynamic and intermediate impact types of loading, and the fatigue life of energetic material is poorly understood. This work presents the effect of thermal and mechanical fatigue loading on the energic material response.

Experiment

Materials and methods

Materials and specimen preparation

This study uses polymer-bonded sugar (PBS) as a material system. Sugar has been widely used as a mechanical simulant of high-melting explosives (HMX). The specimens were prepared by cold pressing of prepreg sugar in a binder and plasticizer [16]. Cylindrical rods were first cold pressed, and cubic specimens were machined. Specimens were prepared carefully for digital image correlation measurement, as discussed in detail [16].

Experiment

A set of specimens was first subjected to cyclic loading, and the damage as a function of the loading cycles was captured. Each specimen was loaded under compressive cyclic loading at 60 % of the yield stress, and the moduli of the specimen as specified loading cycle was measured. The elastic modulus of specimens as a function of loading cycles tested at different loading frequencies is captured. A similar procedure was followed in the temperature cycle to measure the specimen's thermal fatigue damage as a function of the thermal cycle. A set of specimens with different damage as a function of loading and thermal cycle is produced. A fractography analysis of the specimen based on optical microscopy is used to distinguish different types of damage.

Once specimens were subjected to fatigue loading and sustained damage, a Hopkinson bar was used to test the specimens further. A range of specimens with different damage have been loaded under dynamic loading and were compared with virgin materials. The Hokinon bar was modified as a direct-impact gas gun to load the samples with a defined strain and strain rate loading. As shown in the figure, the specimen was attached to the transmitter bar, and a projectile was applied directly to the specimen.

The experiments utilize an ultrahigh-speed camera with microscopic and macro lenses, and a digital image correlation (DIC) technique in Fig. 1. The loading rate and the imaging frame were well synchronized to capture sufficient images during deformation. The digital image correlation data has been used to identify the damage initiation and programming and serve as input to the computational effort.

Computational

To capture the simulations of the crystal-binder interaction, the research group has developed a Material Point Method model to capture the therm-mechanical contact and the potential de-lamination of the crystal.

In the MPM model, the material point flows in the Eulerian background grid, enabling one to overcome the mesh distortion issue of the Lagrangian finite element while retaining the Lagrangian description of the history-dependent material behaviors (see Figure 2).

The modeling focuses on capturing the PBX behaviors by modeling the large-deformation behaviors of the binder, crystals, and the binder-crystal interface. We idealize the polymer binder as a Neo-Hookean material with a viscous response. While the volumetric responses are captured by a modified version of the Birch-Murnahan Equation of State, the dynamic shear modulus is rate-dependent.

The damage is modeled by a cohesive zone model explicitly written for the MPM framework. In the 2D setting, we introduce a cohesive element as massless line segments and assign the interface Lagrangian points. At each time step, the

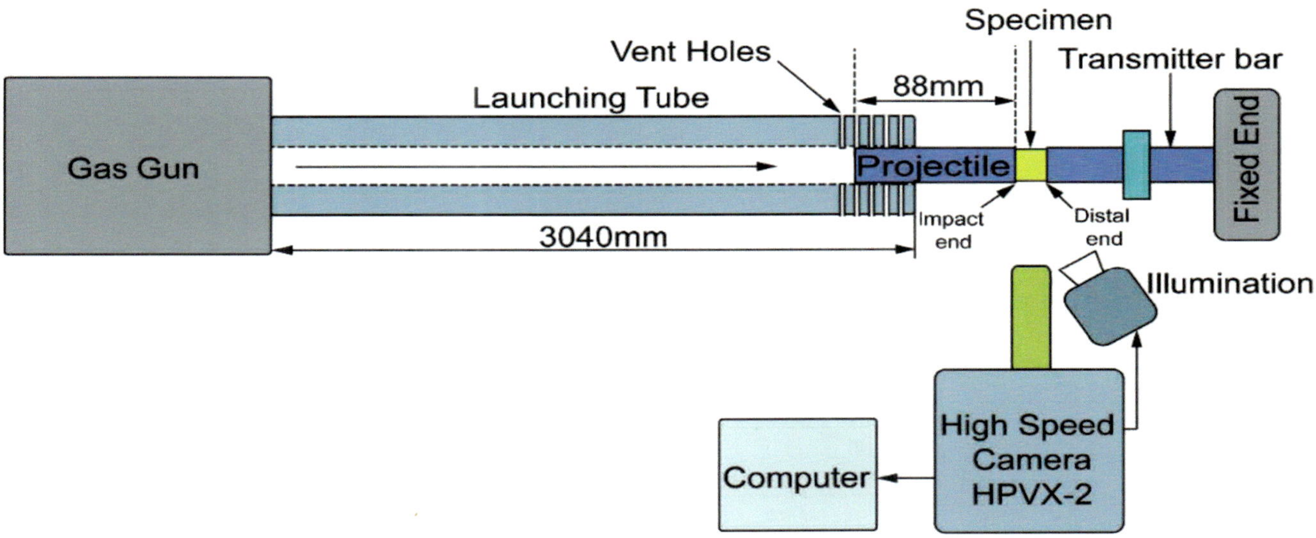

Fig. 1 Experiment setup

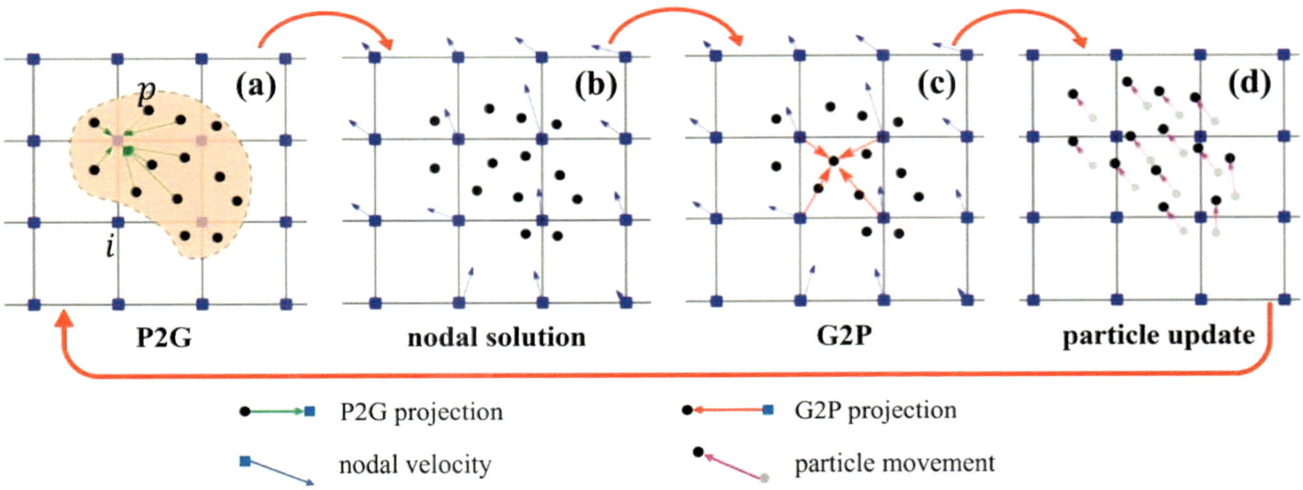

Fig. 2 MPM model description.

displacement between the two interfaces is projected in the Eulerian frame, and the relative displacement between the top and bottom interfaces is updated.

The cohesive zone model takes account of the mixed-mode contributions where an effective displacement is calculated as a weighted sum of the normal and tangential modes of deformation. Temperature dependence on the traction separation law is also captured (see Figure 3).

To capture the potential fracture triggered by the contacts of crystals, frictional responses under compression are also considered, and a contact algorithm is introduced so that the resultant hydrocode can capture the hot spot formation influenced by the frictional heating.

Results

A preliminary study on damage propagation was conducted by subjecting the damaged specimen to quasi-static and dynamic loading conditions. The microstructure image taken during the test indicates that the failure mechanism under quasi-static and dynamic loading differs. A complete fracture of the crystals has been observed in the quasi-static experiments. It is

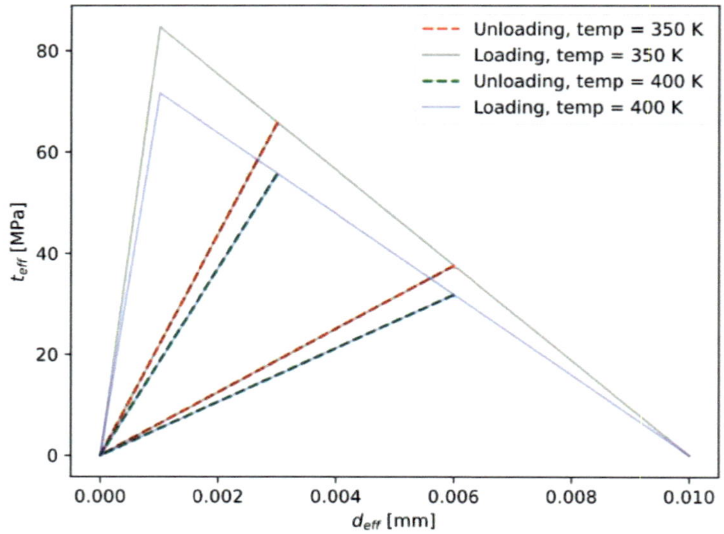

Fig. 3 Temperature-dependent traction separation law

more like a tensile fracture in the crystal and is expected from brittle material like sugar. On the other hand, the dynamic result shows a localized failure where the small crystal is squeezed between two relatively larger crystals. However, this localization is not everywhere but focused on small regions, usually where there is a sharp contact between crystals.

We also observed the high speed of the volume fraction of the crystal. The higher volume fraction material shows more heterogeneity and localization failure.

Conclusion

We have developed an experimental method to capture the effect of thermal and mechanical fatigue loading on the performance of energetic materials. Damage propagation as a function of the loading cycle has been studied; the damaged samples were tested under dynamic loading conditions and were compared with virgin materials. The generative AI enabled us to augment experimental data with synthetic data to meet the demand for more precise modeling.

Acknowledgments The financial support of the Air Force Office of Scientific Research (AFOSR) under Grant number FA9550-22-1-0310 is gratefully acknowledged.

References

1. Zhong, K. and Zhang, C., 2024. Review of the decomposition and energy release mechanisms of novel energetic materials. *Chemical Engineering Journal*, *483*, p.149202.
2. Li, H., Zhang, L., Petrutik, N., Wang, K., Ma, Q., Shem-Tov, D., Zhao, F. and Gozin, M., 2019. Molecular and crystal features of thermostable energetic materials: guidelines for the architecture of "bridged" compounds. *ACS Central Science*, *6*(**??**), pp.54-75.
3. Yang, J., Lu, Z., Zhou, X., Sun, Z., Hu, Y., Zhang, T., Wu, C., Zhang, G. and Jiang, W., 2023. Current self-healing binders for energetic composite material applications. *Molecules*, *28*(**??**), p.428.
4. Ravindran, S., Gupta, V. and Kidane, A., 2023. Mechanism of Hotspot Formation in Energetic Materials Under Mild Impact Loading. *Journal of Dynamic Behavior of Materials*, *9*(**??**), pp.375-383.
5. Ravindran, S., Tessema, A. and Kidane, A., 2016. Local deformation and failure mechanisms of polymer-bonded energetic materials are subjected to high strain rate loading. J Dyn Behav Mater 2: 146–156 .
6. Ravindran, S., Tessema, A. and Kidane, A., 2017. Multiscale damage evolution in polymer bonded sugar under dynamic loading. *Mechanics of Materials*, *114*, pp.97-106.
7. Barua, A., Horie, Y., Zhou, M., 2012a. Energy localization in HMX-Estane polymer-bonded explosives during impact loading. J Appl Phys 111, 54902.

8. Meng, L., Song, Q.G., Yao, C., Zhang, L. and Pang, S.P., 2024. Chemical reaction mechanisms and models of energetic materials: A perspective. *Energetic Materials Frontiers*.
9. Pandey, S.K. and Roy, K., 2024. Predicting the performance and stability parameters of energetic materials (EMs) using a machine learning-based q-RASPR approach. *Energy Advances*, *3*(**??**), pp.1293-1306.
10. Wang, X., Wu, Y., Huang, F., Jiao, T., Clifton, R.J., 2016. Mesoscale thermal-mechanical analysis of impacted granular and polymer-bonded explosives. Mechanics of Materials 99, 68–78.
11. Barua, A., Horie, Y., Zhou, M., 2012b. Microstructural level response of HMX–Estane polymer-bonded explosive under effects of transient stress waves, in: Proceedings of the Royal Society of London A: Mathematical, Physical and Engineering Sciences. The Royal Society, pp. 3725–3744.
12. LaBarbera, D.A., Zikry, M.A., 2013. The effects of microstructural defects on hot spot formation in cyclotrimethylenetrinitramine-polychlorotrifluoroethylene energetic aggregates. J Appl Phys 113, 243502.
13. Ravindran, S, Gupta, V., Tessema, A., Kidane, A., 2019. Effect of particle mass fraction on the multiscale dynamic failure behavior of particulate polymer composites. Exp Mech 59, 599–609.
14. Ravindran, Suraj, Tessema, A., Kidane, A., Jordan, J., 2019. Weak-shock wave propagation in polymer-based particulate composites. J Appl Phys 125.
15. Siviour, C.R., Laity, P.R., Proud, W.G., Field, J.E., Porter, D., Church, P.D., Gould, P., Huntingdon-Thresher, W., 2008. High strain rate properties of a polymer-bonded sugar: their dependence on applied and internal constraints. Proceedings of the Royal Society A: Mathematical, Physical and Engineering Sciences 464, 1229–1255. https://doi.org/10.1098/rspa.2007.0214
16. Ravindran, S., Tessema, A., Kidane, A., 2016b. Note: Dynamic meso-scale full field surface deformation measurement of heterogeneous materials. Review of Scientific Instruments 87. https://doi.org/10.1063/1.4944488

Chapter 11
A Variation on the Expanding Ring Tensile Test

Brady Aydelotte, Warren Jones, Nathan Woods, and Colin Merriman

Abstract The expanding ring tensile test is a method for collecting high strain rate tensile data. It offers high strain rates and large deformations. It can be performed in several ways, including the use of laboratory guns, explosives, and electromagnetic expansion. Each of these methods possess advantages and disadvantages.

In this work we investigate using an RP-4 SC detonator to provide energy for expanding ring experiments and an expanding tube experiment. Three different geometries are considered. We explore the effects of different geometry choices and the ability to extract material model parameters from this version of the expanding ring tensile test.

Keywords Tension, Strength, Expanding Ring, Polymer, Detonator

Introduction

The expanding ring tension test is a method for collecting high strain rate tensile data. It can be performed in a number of different ways including explosives [1], [2], [3], [4], [5], electromagnetic expansion [6], [7], [8], exploding wire [9, pp. 105–106], [10], and a version of it is performed using gun impact on an elastomer filler [11], [12]. Some of the benefits of this approach include allowing the experimentalist to access large plastic strains and high strain rates more easily than most variants of the Kolsky bar tension test.

In this work we examined the possibility of using an RP-4 SC detonator [13] to provide the energy for ring or cylinder expansion. It is somewhat analogous to the method employed by [11], [12] in using a projectile to apply pressure to a filler inside a sample tube to cause expansion and fragmentation. We examined three different candidate designs for coupling the RP-4 SC's energy into a sample using a polymer. Then we examined the viability of these tests for optimization-based fitting of constitutive model parameters by performing a sensitivity analysis.

Method

The ALEGRA hydrocode [14] was used to model the different experimental configurations. An Eulerian, 2D axisymmetric mesh was used for the evaluations. Model response was adequately converged at a cell size of 0.12 mm for the RP-4 SC [15] and this cell size was used for all 2D modeling.

The material models used for this study are shown in Table 1. The constants for the models are drawn from ALEGRA's included constitutive library except where specified.

The RP-4 SC model was taken from [15] which was derived from [16]. The RP-4 SC formation was modeled until the jet tip moved 3 charge diameters from the initial location of the front (open) portion of the copper liner. The model was exported as a collection of exodus files containing complete geometry, thermodynamic and kinetic state of the formed jet. The same formed RP-4 SC model was used for all subsequent candidate design simulations.

The first geometry was a plain cylinder (PC) with a polymer insert and steel caps, shown in Figure 1 (a). This was a control geometry, primarily to determine if the ring expansion geometries offer any meaningful improvement in their ability to fit constitutive model parameters. The second geometry was a ring expansion geometry (R1), shown in Figure 1 (b). This is very similar to the PC except the polymer cylinder was made thin relative to its diameter. The last simulated experimental

Brady Aydelotte · Warren Jones · Nathan Woods · Colin Merriman
Idaho National LaboratoryIdaho Falls, ID 83415
e-mail: brady.aydelotte@inl.gov; warren.jones@inl.gov; charles.woods@inl.gov; colin.merriman@inl.gov

Paul Allison et al. (eds.), *Dynamic Behavior of Materials, Vol. 1*, Conference Proceedings
of the Society for Experimental Mechanics Series, https://doi.org/10.13052/9788743808275-11

Table 1 Material models and equations of state used for different materials. JC refers to either the Johnson-Cook strength [17] or failure [18] model depending on the column. SGL refers to the Steinberg-Guinan-Lund constitutive model [14], [19], [20]. JWL refers to the Jones-Wilkins-Lee equation of state [21]. LANL refers to Los Alamos National Laboratories. SNL refers to Sandia National Laboratories.

Component	Equation of State	Constitutive Model	Failure Model
Air	LANL Sesame 5030	–	–
Confinement Disk, Steel	SNL Sesame 2150	JC 2IN RHA with A=875e6 Pa	JC Steel
Confinement Disk, Copper	SNL Sesame 3325	JC Copper	JC Copper
Expansion Cylinder, Copper	SNL Sesame 3325	JC Copper	JC Copper
Expansion Cylinder, Al	SNL Sesame 3700	JC 6061-T6 Al	JC 6061-T6 Al
Polymer Insert (HDPE)	LANL Sesame 97182 Reactant LANL Sesame 97183 Product Arrhenius Reactive Burn	–	–
RP-4 SC Explosive	JWL Composition C-4	–	–
RP-4 SC Liner	SNL Sesame 3325	SGL Copper	Void insertion P<-2.3 GPa
RP-4 SC Outer shell	SNL Sesame 3700	JC 6061-T6 Al	JC 6061-T6 Al
Sample Ring	SNL Sesame 3700	JC 6061-T6 Al	JC 6061-T6 Al
Sensitivity Analysis model	SNL Sesame 3700	JC 6061-T6 Al with A, B, and N as variables	JC 6061-T6 Al
Tamper, Steel	SNL Sesame 2150	JC 2IN RHA with AJ0=875e6 Pa	JC Steel

geometry, a ring expansion geometry, (R2) had a more sophisticated design for the outer cylinder to give the ring a more uniform push.

DAKOTA [22] was used to perform parametric studies to select the final geometries for the different experimental configurations. These will not be addressed in detail; the final dimensions are shown in Figure 1. Once the geometry was fixed, a sensitivity analysis was performed on each geometry. A random sampling approach was generated using a Latin Hypercube Sampling approach. DAKOTA was allowed to select values of the Johnson-Cook strength model parameters A, B, and n (see Equation 1) and the ring or tube velocity history was modeled. The ring velocity was sampled at five different equally spaced times between the initial rise above zero velocity and the first time its velocity again becomes zero. The Spearman rank correlation sensitivity indices were computed [22].

$$\sigma_Y = \left[A + B(\varepsilon^p)^n\right]\left[1 + Cln\left(\varepsilon^*\right)\right]\left[1 - \theta^M\right] \quad (1)$$

Results

The PC geometry is very simple. The parts are amenable to conventional or additive manufacturing. The radial expansion velocity can be controlled by the thickness of the tube, the diameter of the tube, and the thickness of the confinement disks. The radial and axial velocity of the design shown in Figure 1 a, with a 12 mm inner diameter tube with a 1 mm wall thickness, is shown in Figure 2. The maximum radial velocity for this configuration was 184.8 m/s and the maximum axial velocity is -66.5 m/s. This design has several distinct axial velocity maxima or minima in the velocity curve prior to 20 microseconds.

The R1 geometry is very similar to the PC except for the plastic cylinder design. Some parametric studies were conducted examining the effects of different ring thickness (assuming a square sample ring cross section) with different polymer cylinder thicknesses. The polymer cylinder thickness was minimized to see if that would reduce the axial velocity component of the ring. A disc slightly thicker than the ring provided more radial velocity for a given ring size. A slightly undersized polymer disc led to a slightly lower velocity. Equal thickness polymer disc and sample ring yielded adequate velocity. The thicker the ring, the slower for a fixed diameter. Parametric studies determined that a polymer disk thickness equal to the ring thickness resulted in subjectively good ring velocity with little polymer being expelled around the ring to potentially obscure it.

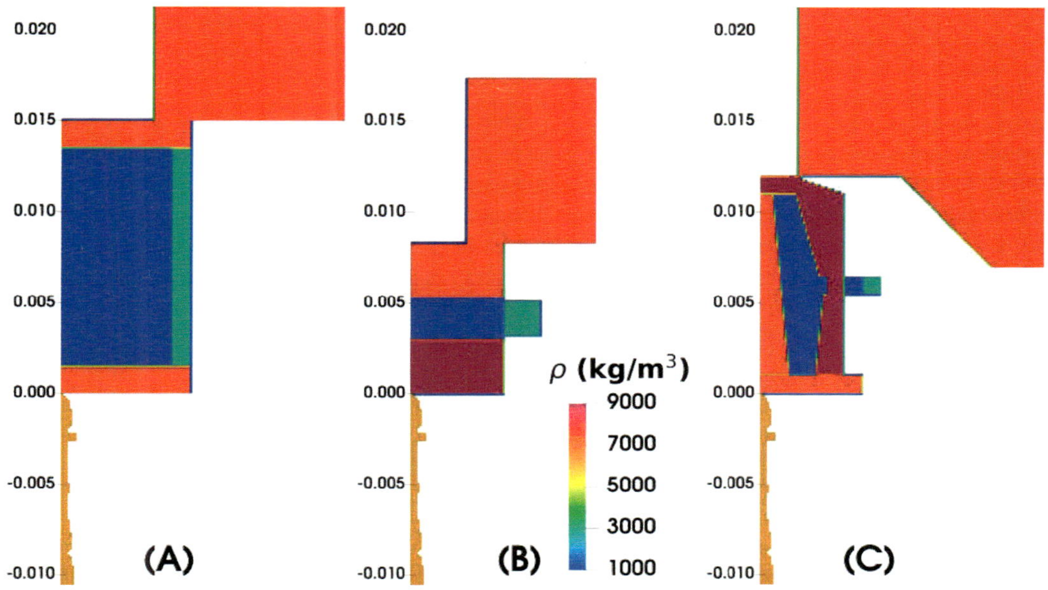

Fig. 1 The three different expansion geometries which were considered; all are radially symmetric about their vertical axes. Dimensions are in meters. Density is plotted for the experimental setups to clarify materials, apart from the shaped charge tip (on the bottom moving up) which is copper colored. The materials used in the design include high density polyethylene, 6061 aluminum, high strength carbon steel, and oxygen free high conductivity copper. (a) The plain cylinder (PC) geometry. (b) The first ring expansion geometry (R1). (c) The second ring expansion geometry (R2).

The radial and axial velocity of the R1 design shown in Figure 1 b, with a 10 mm inner diameter sample ring with a 2 mm wall thickness, is shown in Figure 2. This design exhibits a maximum velocity of 182.5 m/s in this configuration and a maximum axial velocity of -19.7 m/s. The largest axial velocities appear after 20 microseconds.

The R2 ring expansion geometry had a more sophisticated design for the outer cylinder. This design was intended to give the sample ring a more uniform push by shaping the copper buffer, outer tube, and polymer insert. The radial and axial velocity of the R2 design shown in Figure 1 b, with a 11 mm inner diameter sample ring with a 1 mm wall thickness, is

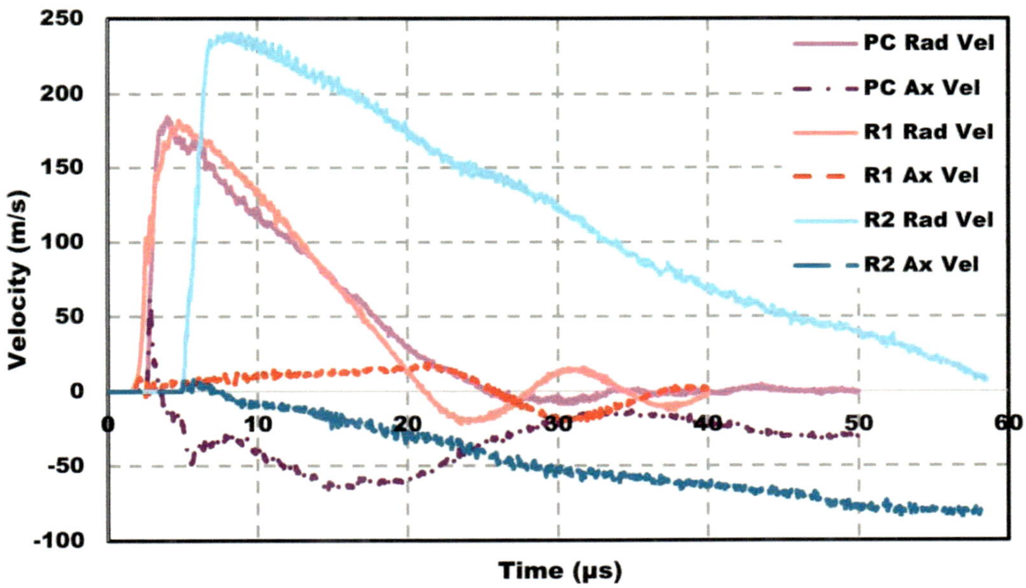

Fig. 2 Plot of radial velocities (the experimental quantity of interest) and the axial velocity (generally undesirable) for the PC, R1, and R2 configurations.

shown in Figure 2. The peak velocity was the highest, of the three designs we considered, at 240.8 m/s with a longer duration of radial velocity greater than zero. The axial velocity was also the highest. However, the maximum occurred at the end of the simulation time window, near 57.5 microseconds.

The sensitivity analysis performed using DAKOTA generated results which are summarized in Figure 3. The absolute value of the sensitivity indices for all the velocity history data points were averaged for Johnson-Cook parameters A, B, and n for each design. The plain cylinder geometry exhibited good sensitivity for A, B, and N, with sensitivity indices which were on average higher than the other two designs. The R2 design averaged the lowest sensitivity indices.

In this case, a good experimental design should exhibit a high degree of sensitivity in the sample velocity to changes in the A, B, and n parameters and all these designs appear to exhibit reasonable sensitivity to the Johnson-Cook parameters in question. It should be possible to use any of the designs for optimization-based material model parameterization. The tube-based design, PC, appears to be the best choice for optimization-based material model fitting of the three considered here.

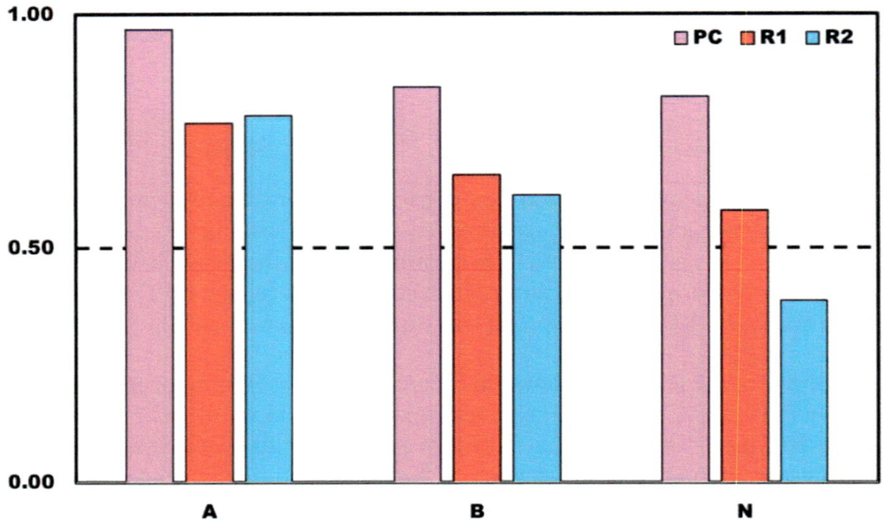

Fig. 3 Average magnitudes of the sensitivity indices for the PC, R1, and R2 for the Johnson-Cook parameters A, B, and n.

There are advantages to ring-based experimental geometries. Less sample material is required, which can be an advantage when the sample material is costly, toxic, or radioactive. It is also useful that the ring velocity data can be used to determine stress and strain directly. The R1 design appears to offer an acceptable compromise of properties which would allow it to be used for either model fitting or direct constitutive data extraction.

Conclusion

We examined three different sample tensile expansion geometries which used an RP-4 SC detonator as an energy source: a design based on a simple sample tube and two different ring-based sample designs. The three designs were evaluated by modeling using the ALEGRA hydrocode and running sensitivity analyses using DAKOTA to assess whether they would be suitable for determining the A, B, and n parameters of the Johnson-Cook constitutive model.

It appears that the tube-based design offers greater sensitivity to the Johnson-Cook constitutive parameters in question and may be a better choice for optimization-based fitting of a constitutive model. However, the ring-based designs should allow direct extraction of material constitutive model parameters more readily than the tube-based design. The first of the ring-based designs, based on using a thin disc of polymer as the driver material for the sample ring, appears to be an acceptable compromise for either model fitting or direct extraction of stress strain data.

Acknowledgments This research made use of Idaho National Laboratory's High Performance Computing systems located at the Collaborative Computing Center and supported by the Office of Nuclear Energy of the U.S. Department of Energy and the Nuclear Science User Facilities under Contract No. DE-AC07-05ID14517.

References

1. P. Johnson, B. Stein, and R. Davis, "Measurement of dynamic plastic flow properties under uniform stress," presented at the symposium on dynamic behavior of materials, ASTM International, 1963.
2. C. Hoggatt and R. Recht, "Stress-strain data obtained at high rates using an expanding ring: Investigation indicates that dynamic symmetrical free expansion of thin rings offers a valid means for obtaining uniaxial tensile stress-strain relationships at high strain rates," Exp. Mech., vol. 9, pp. 441–448, 1969.
3. R. Warnes, T. Duffey, R. Karpp, and A. Carden, "An improved technique for determining dynamic material properties using the expanding ring," Shock Waves High-Strain-Rate Phenom. Met. Concepts Appl., pp. 23–36, 1981.
4. B. Aydelotte, "Analysis of the explosively driven expanding ring tension test," presented at the Dynamic Behavior of Materials, Volume 1: Proceedings of the 2020 Annual Conference on Experimental and Applied Mechanics, Springer, 2021, pp. 73–77.
5. B. Aydelotte, P. Hibner, B. D. Brown, W. Fuger, C. Angell, and W. Jones, "Additional Study of the Explosively Driven Expanding Ring Tension Test," in Dynamic Behavior of Materials, Volume 1, S. Mates and V. Eliasson, Eds., Cham: Springer International Publishing, 2022, pp. 83–87.
6. F. I. Niordson, "A unit for testing materials at high strain rates," Exp. Mech., vol. 5, no. 1, pp. 29–32, Jan. 1965, doi: 10.1007/BF02320901.
7. D. Grady and D. Benson, "Fragmentation of metal rings by electromagnetic loading," Exp. Mech., vol. 23, no. 4, pp. 393–400, Dec. 1983, doi: 10.1007/BF02330054.
8. H. Zhang and K. Ravi-Chandar, "On the dynamics of necking and fragmentation – I. Real-time and post-mortem observations in Al 6061-O," Int. J. Fract., vol. 142, no. 3, pp. 183–217, 2006, doi: 10.1007/s10704-006-9024-7.
9. B. Aydelotte, "Fragmentation and reaction of structural energetic materials," Goergia Institute of Technology, Atlanta, GA, 2013. [Online]. Available: http://hdl.handle.net/1853/50253
10. A. Fedorenko, Y. Sudenkov, S. Konev, and I. Sergeichev, "Exploding wire method for the characterization of dynamic tensile strength of composite materials," Int. J. Impact Eng., vol. 180, p. 104704, Oct. 2023, doi: 10.1016/j.ijimpeng.2023.104704.
11. R. E. Winter and H. G. Prestige, "A technique for the measurement of the high strain rate ductility of metals," J. Mater. Sci. Lett., vol. 13, no. 8, pp. 1835–1837, Jan. 1979, doi: 10.1007/BF00548753.
12. T. J. Vogler et al., "Fragmentation of Materials In Expanding Tube Experiments," Int. J. Impact Eng., vol. 29, no. 1–10, pp. 735–746, Dec. 2003, doi: 10.1016/j.ijimpeng.2003.10.020.
13. "RP-4 SC EBW Shaped Charge." RISI, 2020. [Online]. Available: https://www.teledynedefenseelectronics.com/energetics/Documents/RP-4SC_2020.pdf
14. T. J. Fuller et al., "ALEGRA User Manual (Code Version 8.17.5)," Sandia National Laboratories, Albuquerque, NM, Oct. 2024.
15. B. Aydelotte, C. Merriman, and W. Jones, "On Depth of Penetration Testing," in Proceedings of the 2024 SEM Annual Conference & Exposition on Experimental and Applied Mechanics, Feb. 2024.
16. D. R. Scheffler, M. S. Burkins, and W. P. Walters, "Characterization of Jets from Exploding Bridge Wire Detonators," Army Res. Lab Aberd. Proving Ground Md Weapons Mater. Res. Dir. ARL-TR, vol. 3518, 2005.
17. G. R. Johnson and W. H. Cook, "A constitutive model and data for metals subjected to large strains, high strain rates, and high temperatures," in Proceedings of the 7th International Symposium on Ballistics, The Hague, Netherlands, 1983, pp. 541–547.
18. G. R. Johnson and W. H. Cook, "Fracture characteristics of three metals subjected to various strains, strain rates, temperatures and pressures," Eng. Fract. Mech., vol. 21, no. 1, pp. 31–48, 1985, doi: doi: DOI: 10.1016/0013-7944(85)90052-9.
19. D. J. Steinberg, S. G. Cochran, and M. W. Guinan, "A constitutive model for metals applicable at high-strain rate," J. Appl. Phys., vol. 51, no. 3, pp. 1498–1504, 1980.
20. D. J. Steinberg and C. M. Lund, "A constitutive model for strain rates from 10−4 to 106 s−1," J. Appl. Phys., vol. 65, no. 4, p. 1528, 1989, doi: 10.1063/1.342968.
21. E. L. Lee, H. C. Hornig, and J. W. Kury, "Adiabatic Expansion Of High Explosive Detonation Products," United States, 1968. doi: 10.2172/4783904.
22. B. M. Adams et al., "Dakota, A Multilevel Parallel Object-Oriented Framework for Design Optimization, Parameter Estimation, Uncertainty Quantification, and Sensitivity Analysis (V.6.16 User's Manual)," Sandia National Lab. (SNL-NM), Albuquerque, NM (United States), SAND2022-6171, Nov. 2021. doi: 10.2172/1868142.

Chapter 12
Characterization of the Human Eardrum Transfer Functions Under Moderate and High-Intensity Sound Levels by High-Speed 3D-DIC

Jonathan Oliveira Luiz, John Rosowski, Yoshiaki Inuzuka, Cosme Furlong, and Jeffrey Cheng

Abstract Characterizing the damage mechanics of the human middle ear under different acoustic excitation levels is still under investigation. Limited research exists on the non-linear behavior of the human tympanic membrane (TM), commonly known as the eardrum. Specifically, when it's exposed to high-intensity sounds (i.e., >130 dB SPL), such as loud music and blasts, despite its accepted linear mechanics under lower sound pressures (i.e., <120 dB SPL) typical of everyday environments. Our research aims to characterize the transition of TM responses from linear to non-linear behaviors with increasing sound levels, identifying sound thresholds at which non-linear behaviors emerge. The methodology under development integrates High-Speed Three-Dimensional Digital Image Correlation (HS 3D-DIC), optimized for small field-of-view (FOV), within a custom-built acoustic system. It uses a loudspeaker to generate moderate-level impulsive sounds (100–160 dB SPL) and a shock tube to produce high-intensity blasts (150–180 dB SPL), enabling a TM response evaluation across a wide range of impulsive sound levels. Using two high-speed cameras at up to 100,000 frames per second, HS 3D-DIC captures TM motions ranging from sub-micrometer to millimeter scales, while high-speed pressure sensors characterize pressure profiles near the TM. Moreover, sample preparation effects are assessed. Transfer functions (TFs) for over 60,000 points on the TM surface are computed by normalizing displacements with sound pressures using Fourier analysis, facilitating comparisons of TM mode shapes at different pressure levels. Initial results reveal linear TM TFs at moderate levels consistent with literature, while non-linear TM TFs at high intensities are still under investigation. This work is expected to enhance knowledge of TM mechanics at different levels and assist in the development of grafting materials and methods to repair the TM.

Keywords Tympanic membrane · Acoustical Transfer Functions · Nonlinearity · High-speed 3D-DIC · Shock tube · Ear injury

Introduction

The human middle ear is an air-filled cavity that includes the eardrum (tympanic membrane, TM) and three tiny bones known as ossicles (malleus, incus, stapes). The TM generates complex, highly frequency-dependent, and multifaceted patterns of vibrations when sound waves hit it [1, 2, 3]. The ossicles couple the resulting sound energy from the TM to the inner ear,

Jonathan Oliveira Luiz · Cosme Furlong
Center for Holographic Studies and Laser micro-mechaTronics (CHSLT), Worcester Polytechnic Institute, Worcester, MA 01609
e-mail: joluiz@wpi.edu; cfurlong@wpi.edu

Jonathan Oliveira Luiz · Cosme Furlong
Mechanical and Materials Engineering DepartmentWorcester Polytechnic Institute, Worcester, MA 01609
e-mail: joluiz@wpi.edu; cfurlong@wpi.edu

John Rosowski · Yoshiaki Inuzuka · Jeffrey Cheng
Eaton-Peabody Laboratories, Massachusetts Eye and Ear, Boston, MA 02114
e-mail: john_rosowski@meei.harvard.edu; yinuzuka@meei.harvard.edu; tao_cheng@meei.harvard.edu

John Rosowski · Yoshiaki Inuzuka · Jeffrey Cheng
Department of Otolaryngology-Head and Neck Surgery Harvard Medical School, Boston, MA 02114
e-mail: john_rosowski@meei.harvard.edu; yinuzuka@meei.harvard.edu; tao_cheng@meei.harvard.edu

Paul Allison et al. (eds.), *Dynamic Behavior of Materials, Vol. 1*, Conference Proceedings
of the Society for Experimental Mechanics Series, https://doi.org/10.13052/9788743808275-12

where sensory hair cells stimulate nerve signals to the brain. The middle ear transfer function (TF) describes how sound pressure at the TM is transformed into motion at a downstream location, such as the stapes or umbo, characterizing the frequency-dependent input-output relationship. This function quantifies how the middle ear amplifies, attenuates, or phase-shifts sound energy across different frequencies, providing insight into normal and pathological hearing function [4].

It is well accepted that at low to moderate levels ($<\sim$120 dB SPL), the middle ear operates linearly, with resulting displacements ranging in the nanometer to micrometer range [5]. This means the input exerted into the TM (sound pressure) is proportional to the output (motions). This way, for transfer functions computed in the linear range, in which motions are normalized by pressure, it is expected that results are always consistent. At higher sound levels (above $\sim$120 dB SPL), however, the displacements grow to the micrometer to millimeter range [1], and nonlinearities emerge, altering the resulting transfer functions. The exact threshold at which nonlinear behavior begins and the mechanisms driving these nonlinearities remain unclear, hindering accurate middle ear models, TM repair methods, and effective hearing protection design [5].

Nonlinearities arise when the relationship between input and output variables, such as force and displacement, deviates from proportionality. They can be broadly classified into geometric and material nonlinearities. Geometric nonlinearity occurs when large deformations significantly alter the structure's shape, causing stiffness to become a function of displacement. This effect becomes relevant when strain exceeds typical thresholds, around 5%, leading to changes in structural behavior [6]. Material nonlinearity, on the other hand, occurs when a material's stress-strain relationship is not constant, often due to viscoelasticity, plasticity, or hyperelasticity [7]. Many biological and engineering materials exhibit both types of nonlinearities, affecting system behavior under varying loads and excitation conditions. Accounting for these effects improves the accuracy of models describing mechanical responses in complex structures, such as the middle ear.

In this work, we compare the behavior of the human TM at moderate (linear response) and high-intensity (nonlinear response) sound levels. We aim to investigate how the TM's vibration response changes as sound pressure increases, focusing on the onset and progression of nonlinearities. Examining middle ear nonlinearities requires assessing both geometrical and mechanical nonlinearities, as well as rate-dependent behaviors, across varying sound levels and excitation rates. Nonlinearities in the middle ear have been investigated in both animal models and humans [8, 9]. However, further clarification is needed for the human middle ear, especially nonlinearities across different sound levels using full-field techniques to describe the TM complex behavior and determine the nonlinearity threshold more accurately.

By applying a methodology that uses a validated integrated system, we capture the complex motion of the tympanic membrane across various sound levels and excitation rates. Through Fourier analysis, we compute the corresponding transfer functions for different conditions and compare the differences for varying excitations. Initial results give new insights into middle ear nonlinear behavior and will be used for future evaluation and analysis.

Methods

In this study, we present a methodology specifically developed to investigate nonlinearities in the human middle ear [1, 10]. The integrated experimental setup allows for two excitation sources: a loudspeaker generating acoustic clicks for low-to-moderate impulse-like pressure waveforms and a pressure-adjustable, fully characterized, repeatable shock tube producing blast waves for high-intensity sound levels [1, 10, 11]. These sources are synchronized with high-speed measurement techniques.

To capture the complex full-field response of the tympanic membrane (TM), high-speed 3D Digital Image Correlation (3D-DIC) provides full-field 3D motion data across the entire TM. A differential Laser Doppler Vibrometer (LDV) is used for two purposes: 1) validating DIC measurements and 2) measuring specific points of interest in the middle ear (e.g., umbo or stapes). The high-density data from DIC enables a detailed analysis of TM kinetics and kinematics. Additionally, a high-speed Schlieren imaging system visualizes the fluid-solid interaction between the temporal bone and the shock wave [12]. High-frequency pressure sensors measure the pressure near the sample, completing pressure characterization. A representation of the integrated system is shown in Fig. 1. An insert, shown in Fig. 1 b, details the temporal bone (a complex bone within the skull base that contains, between other structures, the middle ear) prepared for an experiment.

The application of 3D-DIC in this context presents several challenges: (1) a small field of view (FOV), as the human tympanic membrane is only $\sim$9 mm in diameter; (2) a long working distance, since high-speed cameras must be positioned away from the shock tube to prevent damage; (3) the need for high-speed measurements to accurately capture rapid TM motion; and (4) minimal sample preparation, as the TM is a thin membrane ($<$150 μm), and a heavy paint layer would alter its natural response. Challenges 1, 2, and 3 are addressed using specialized equipment, including high-magnification lenses and high-speed cameras, enabling measurement speeds up to 100k FPS. This system was previously tested and validated [1, 11, 12, 13]. Given the specialized nature of this 3D-DIC application, we validated our measurements under two conditions: moderate sound levels and high-intensity pressure levels. Full-field 3D-DIC data were compared against

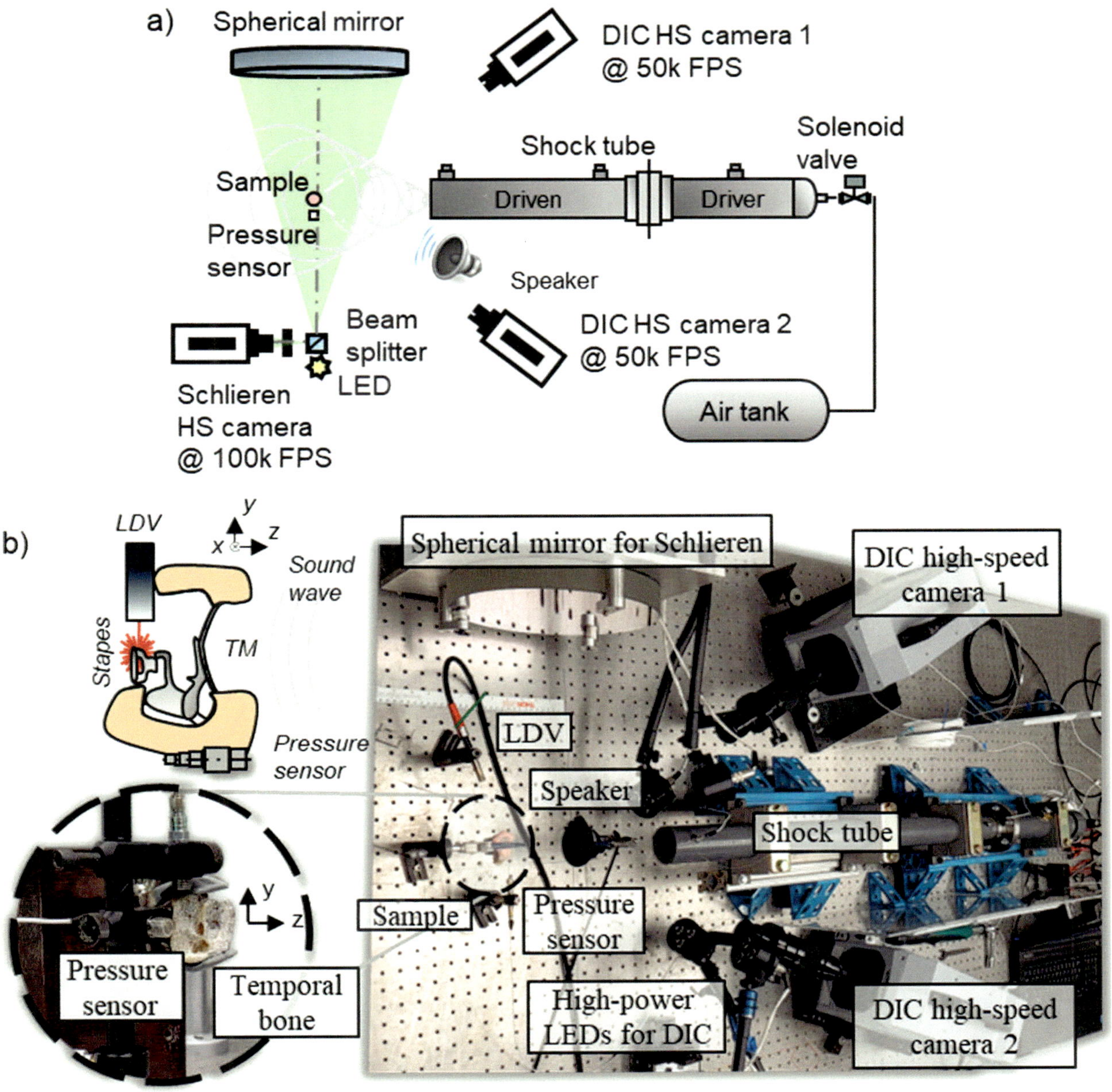

Fig. 1 a) Schematics of the integrated experimental setup; b) Realization of the integrated system with an inset of a cadaveric temporal bone under test.

LDV measurements at the center of a 10 mm artificial membrane to assess accuracy, as depicted in Fig. 2. Despite the event lasting only a few milliseconds, 3D-DIC and LDV measurements for both click and blast exposures showed an average error of less than 1%, confirming the accuracy of 3D-DIC at moderate and high-intensity levels.

To assess challenge number 4, the effects of painting on thin membranes with responses similar to the human TM were addressed. Tunable thin membranes capable of maintaining a consistent response were designed and fabricated, as shown in Fig. 3 a. These membranes are made of latex, with a diameter of 10 mm and a thickness of $\sim$100 μm. The painting process involved using an airbrush to apply different coatings to the artificial sample. The response to acoustic excitation was compared before and after painting to determine if the response remained unchanged. For this, transfer functions at the center of the sample were computed using an LDV. Various coating materials were tested, as shown in Fig. 3 a. Typically, for DIC painting, a white background is first applied, followed by black speckles to achieve better contrast [14]. In this study, however, we tested a less conventional method: only white speckles were applied to minimize the added mass to the

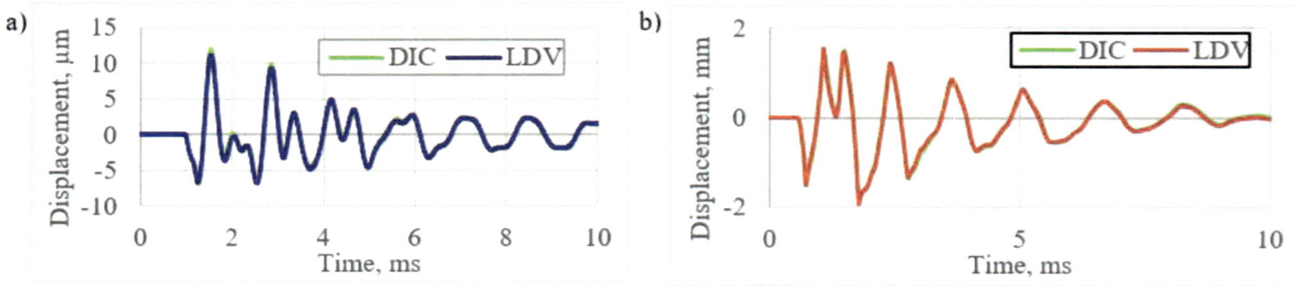

Fig. 2 Validation at two pressure levels: a) Acoustic click, ∼35 Pa (125 dB SPL); b) Blast exposure, 20 kPa (180 dB SPL).

sample and avoid altering the membrane stiffness, which could be caused by a uniform background paint, thus reducing any potential impacts the coating could have on the thin membrane's natural response.

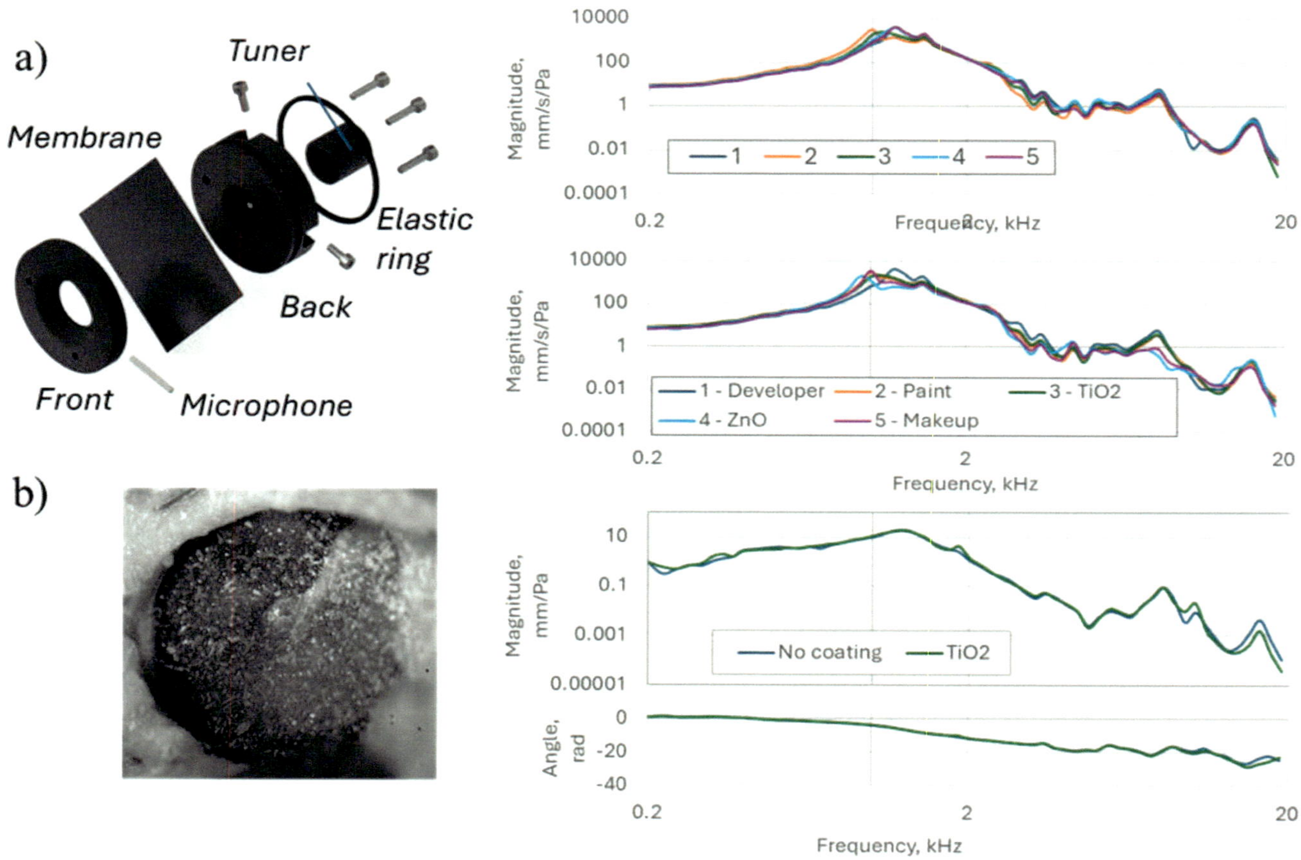

Fig. 3 Painting evaluation: a) Tunable thin membrane with bracket designed for consistent response; b) Test on a real TM with TiO_2 speckles.

The coating that yielded the best results, TiO2 speckles, was applied to a real TM, as shown in Fig. 3 b. TF measurements taken at the umbo, before and after sample preparation, show no significant change in the tympanic membrane's response, confirming that the sample preparation method does not alter its characteristics. Additionally, TiO2 speckles were successfully tracked by the 3D-DIC algorithm while maintaining a minimally invasive preparation. All DIC computations in this work were performed using VIC-3D software from Correlated Solutions [15].

In a previous study, we were unable to compute TFs for blast experiments due to issues such as pressure sampling acquisition and vortex rings, which caused pressure waveforms to deviate from an impulse-like response [1]. In this work, we adopted a sampling rate of 200 kHz for pressure acquisition and eliminated vortex rings by positioning the shock tube off-axis, as shown in Fig. 4. While the primary shock wave remains spherical and interacts with the sample regardless of the shock tube's alignment with the specimen, the vortex ring travels along the shock tube's longitudinal axis, preventing interaction with the tested temporal bone.

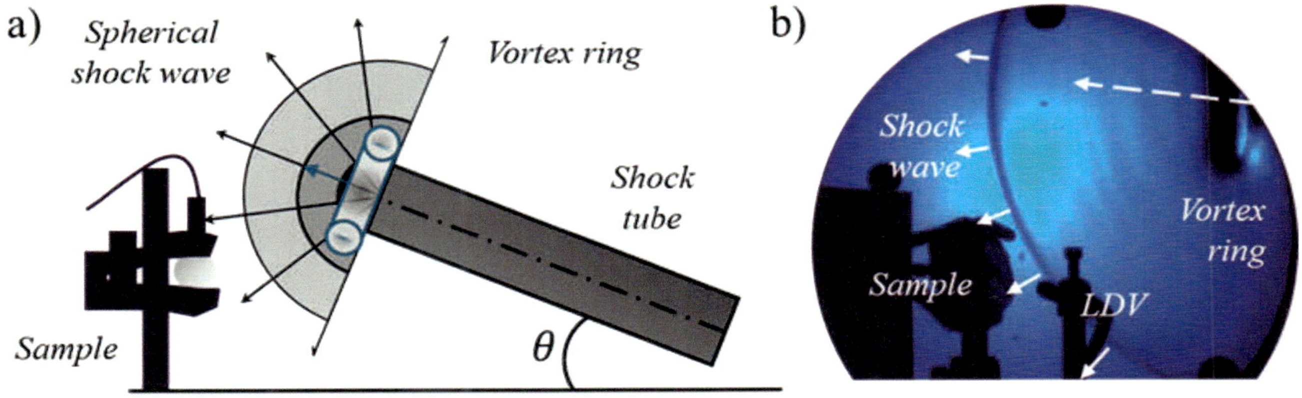

Fig. 4 Impulse-like blast exposure: a) Representation of shock tube off-axis configuration; b) Schlieren image showing fluid-solid interaction between shock wave and temporal bone, while vortex ring deviates from the tested sample.

The tested samples consist of cadaveric human temporal bones, carefully prepared for these experiments. Transfer functions for different excitation levels across the entire surface of the TM were computed using the equation

$$TF\left(x, y, f\right) = \frac{FFT\left(w\left(x, y, t\right)\right)}{FFT\left(p(t)\right)}, \tag{1}$$

where FFT represents a fast Fourier transform, w is displacement along z-direction, and p is the pressure measured near the sample. Additionally, impulse responses (IRs) are computed by

$$IR\left(x, y, t\right) = IFFT\left(TF\left(x, y, f\right)\right). \tag{2}$$

Finally, as an additional method to assess potential nonlinearities, we compute the ratio of volume displacement and pressure energy flux accumulated over time, as in

$$Ratio\left(t\right) = \frac{E^*}{V} = \frac{\frac{1}{\rho c}\int_0^T p^2\left(t\right)dt}{\int_0^T \int_a^b \int_c^d w\left(x, y\right)dydxdt}. \tag{3}$$

Results

As a representative result, we present time-domain measurements of the tympanic membrane (TM) using high-speed 3D-DIC at two excitation levels: an acoustic click (2000 Pa, 160 dB SPL) in Fig. 5 a and a blast (15 kPa, 177.5 dB SPL) in Fig. 5 b. Displacement data are shown along the z-direction at three points of interest (POIs), illustrating the TM's response under both loading conditions. The click is expected to be closer to the linear range of the TM, while the blast falls within the nonlinear regime. Specific instances are highlighted to demonstrate key deformation behaviors, with detailed motion decomposition at time instance t_3, as shown in Fig. 5 c and Fig. 5 d.

The data shown in Fig. 5 is used to compute the transfer function, following Eq. 1. The results are presented in Fig. 6. For the identified dominant frequencies, the corresponding mode shapes are computed at both moderate and high-intensity sound levels. The reduced magnitude and altered shape of the blast-induced transfer functions indicate a significant frequency-dependent compressive nonlinearity.

Impulse responses characterize the time-domain behavior of the TM by describing its motion following a brief acoustic stimulus. They provide insights into how the TM and connected structures respond over time, revealing details about damping, resonance, and energy transmission in both normal and pathological conditions. The corresponding impulse responses were computed using Eq. 2, with the results shown in Fig. 7. Key features of the impulse responses are examined to identify

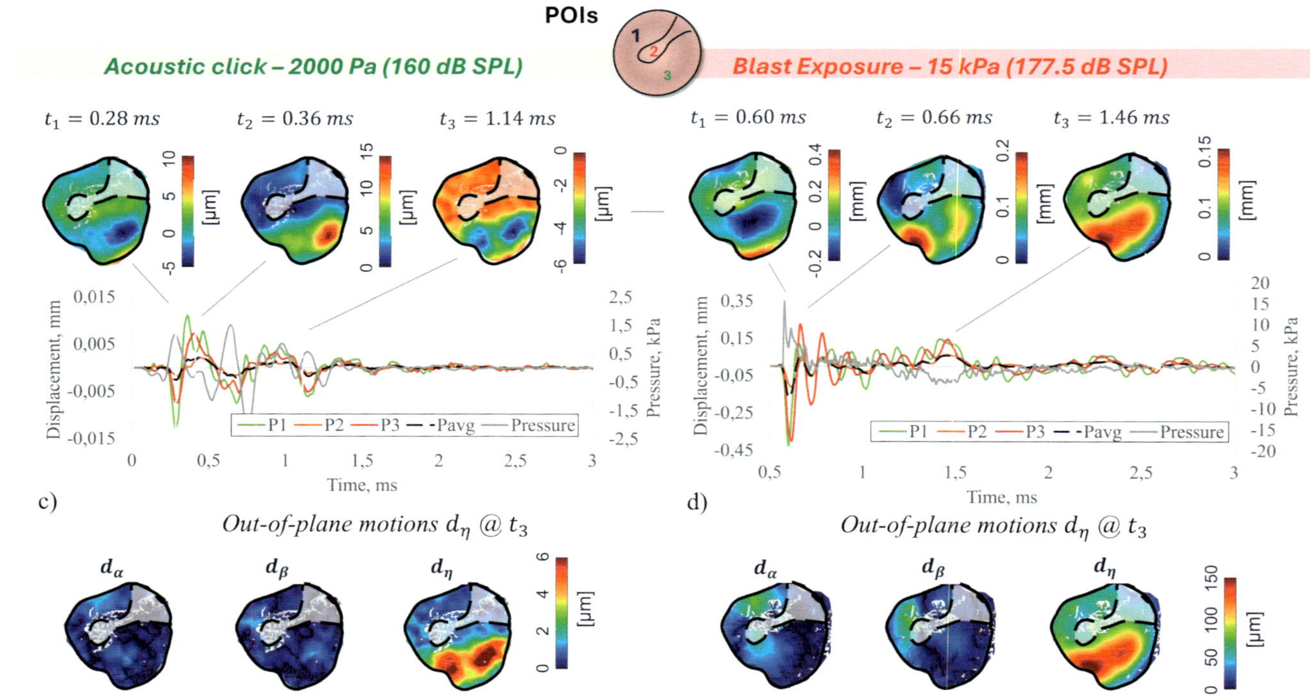

Fig. 5 Time-domain measurement of a temporal bone exposed to: a) Acoustical click (50 Pa, 128 dB SPL); b) Blast (15 kPa, 177.5 dB SPL). c) and d) show motion decomposition for the corresponding time instances at t_3.

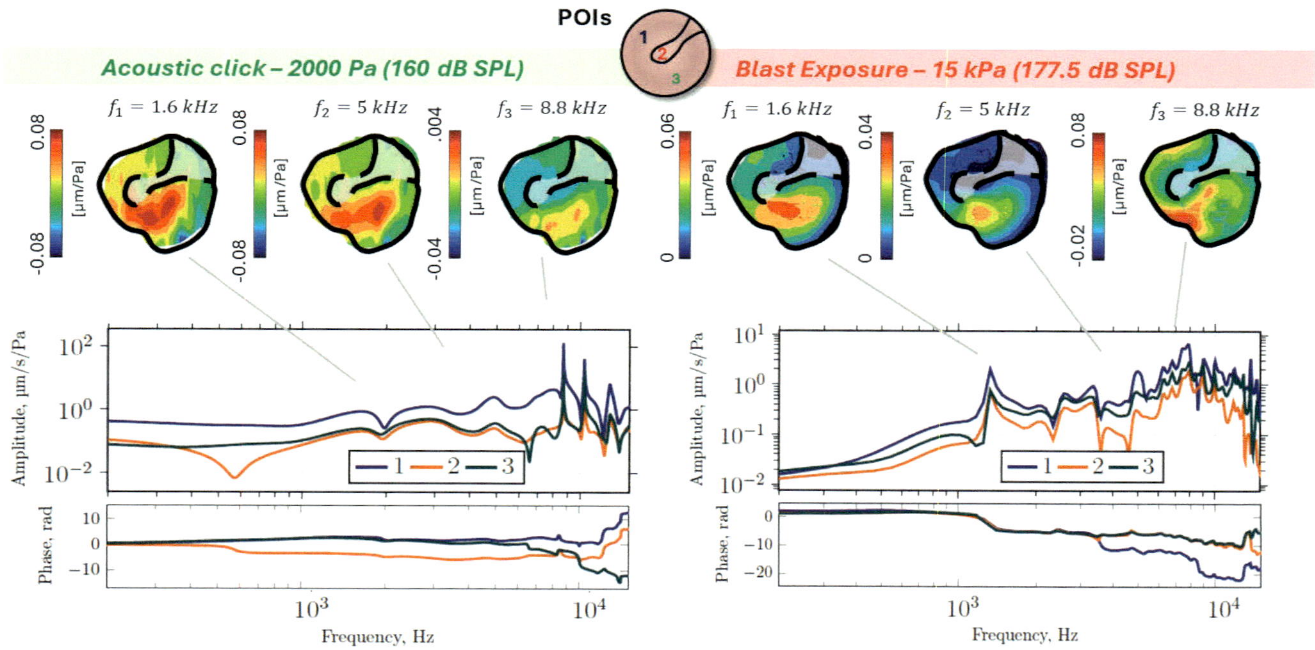

Fig. 6 Transfer functions computed for: a) Acoustical click (2000 Pa, 160 dB SPL); b) Blast (15 kPa, 177.5 dB SPL).

spatial variations across the TM. At high-intensity levels, the reduced magnitude and altered motion patterns further suggest a compressive nonlinearity.

The ratio of volume displacement to pressure energy flux assesses how efficiently acoustic energy is converted into TM motion. Since pressure energy flux represents the input energy and displaced volume reflects the TM's response, discrep-

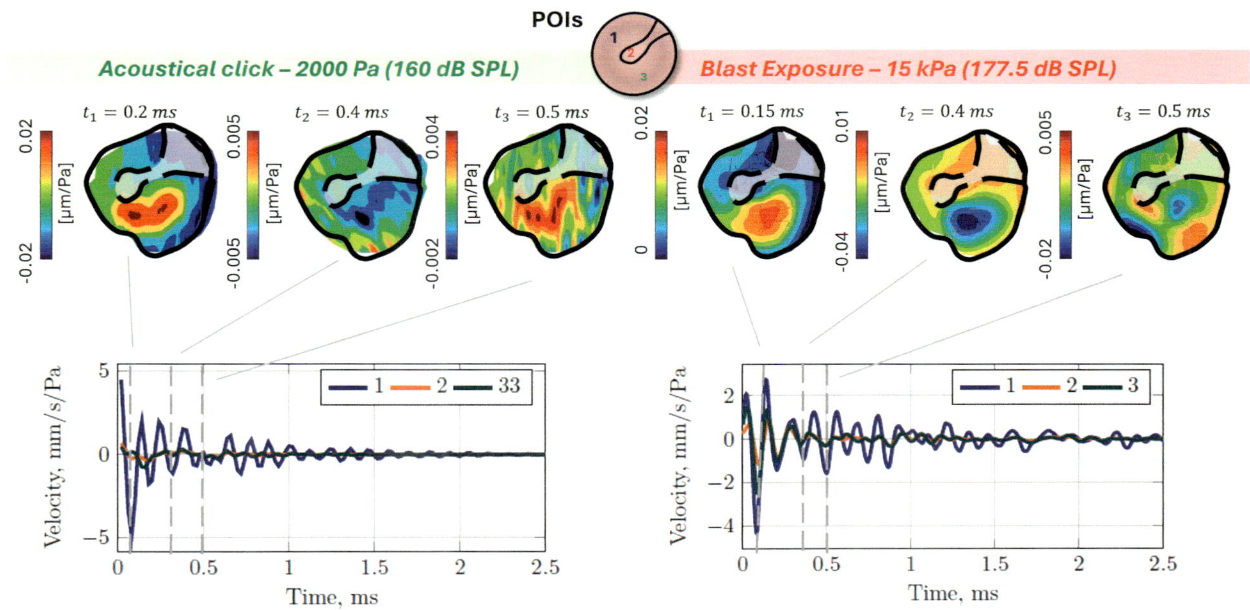

Fig. 7 Impulse responses computed for: a) Acoustical click (2000 Pa, 160 dB SPL); b) Blast (15 kPa, 177.5 dB SPL).

ancies at higher intensities provide further evidence of nonlinearities in the middle ear. Comparing displaced volume and pressure energy flux reveals both kinematic and kinetic effects, offering insight into nonlinearities and energy transfer variations across sound levels. This ratio is calculated for the same cases shown in Fig. 5 using Eq. 3 and is presented in Fig. 8. Deviations from a consistent ratio at moderate and high intensities indicate nonlinear behavior, with the slower growth of Volume/Pressure further suggesting some level of compressive nonlinearity.

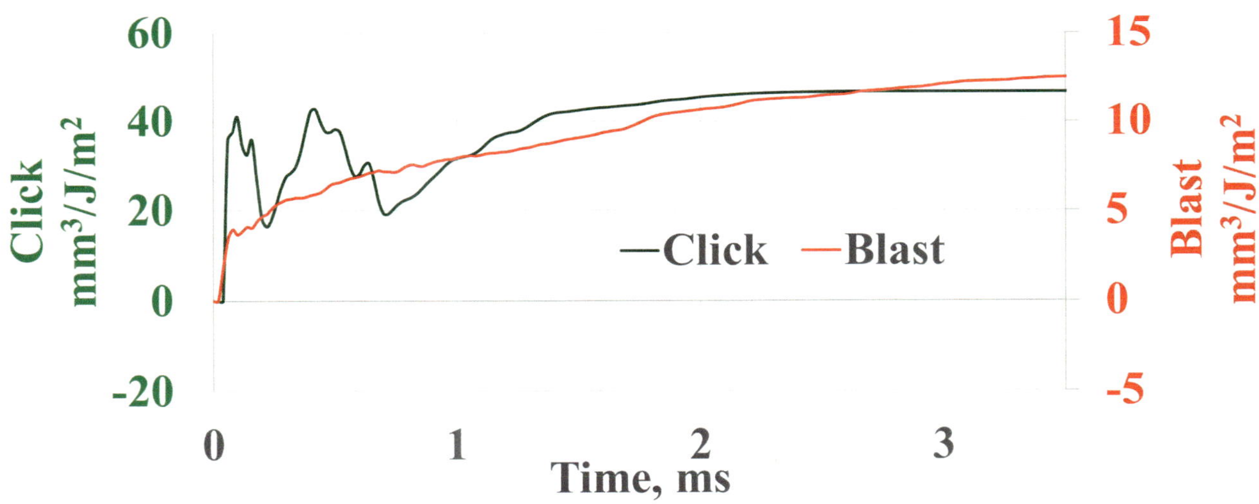

Fig. 8 Volume displacement vs. Pressure energy flux.

Conclusion and Future Work

This study demonstrates the capability of high-speed 3D-DIC to provide detailed spatial and temporal characterization of TM motion at moderate to intense excitation levels, surpassing traditional single-point measurement techniques. The initial results indicate that the middle ear exhibits some level of compressive nonlinearity at high-intensity sound levels, with

TM displacement growing more slowly than pressure increases. Changes in transfer functions, mode shapes, and impulse responses suggest the presence of both geometrical and mechanical nonlinearities.

To further validate these findings, additional samples and excitation levels are needed to confirm observed trends and enhance statistical significance. Future work will focus on expanding the dataset and applying data mining and machine learning models to extract deeper insights into middle ear mechanics. These efforts aim to improve hearing protection design and diagnostic methods. Additionally, the minimal sample preparation required for 3D-DIC measurements suggests potential for in vivo applications, which could enable direct assessments in patients and further advance clinical and research capabilities in auditory science.

Acknowledgments This work has been funded by Award Number R01 DC016079 from National Institutes of Health - National Institute on Deafness and Other Communication Disorders (NIH/NIDCD).

References

1. Oliveira Luiz, J., Alipanahi, A., Rosowski, J., Furlong, C., and Cheng, J. T., "Quantifying real-time dynamic responses and damage mechanics of human tympanic membranes exposed to blast waves," *ASME J of Medical Diagnostics,* 2025. doi:10.1115/1.4067892.
2. Tang, H., Psota, P., Rosowski, J. J., Furlong, C., and Cheng, J. T., "Analyses of the tympanic membrane impulse response measured with high-speed holography," *Hear. Res.*, *410*, p. 108335, 2021. doi:10.1016/j.heares.2021.108335.
3. Rosowski, J. J., "Outer and middle ears," in *Comparative Hearing: Mammals*, R. R. Fay, and A. N. Popper, eds., Springer New York, New York, NY, pp. 172–247, 1994. doi:10.1007/978-1-4612-2700-7_6.
4. Funnell, W. R. J., Maftoon, N., and Decraemer, W. F., "Modeling of middle ear mechanics," pp. 171–210, 2013. doi:10.1007/978-1-4614-6591-1_7.
5. Tang, H., Psota, P., Rosowski, J. J., Furlong, C., and Cheng, J. T., "Ultra-high speed holographic shape and displacement measurements in the hearing sciences," *Light: Adv. Manuf.*, *3*(1), p. 015, 2022. doi:10.37188/lam.2022.015.
6. Liang, J., Luo, H., Yokell, Z., Nakmali, D. U., Gan, R. Z., and Lu, H., "Characterization of the nonlinear elastic behavior of chinchilla tympanic membrane using micro-fringe projection," *Hear. Res.*, *339*, pp. 1–11, 2016. doi:10.1016/j.heares.2016.06.007.
7. Cheng, J. T., Ghanad, I., Remenschneider, A., and Rosowski, J., "The onset of nonlinear growth of middle-ear responses to high-intensity sounds," *Hear. Res.*, *405*, p. 108242, 2021. doi:10.1016/j.heares.2021.108242.
8. Aerts, J. R., and Dirckx, J. J., "Nonlinearity in eardrum vibration as a function of frequency and sound pressure," *Hear. Res.*, *263*(1-2), pp. 26–32, 2010. doi:10.1016/j.heares.2009.12.022.
9. Cheng, J. T., Remenschneider, A., Kozin, E., Furlong, C., and Rosowski, J. J., "Nonlinear response of human middle ear to high level sound," *J. Acoust. Soc. Am.*, 141(5, Suppl.), p. 3897, 2017. doi:10.1121/1.4988760.
10. Alipanahi, A., Oliveira Luiz, Rosowski, J., Furlong, C., Cheng, J., "High-speed three-dimensional-digital image correlation and schlieren imaging integrated with shock tube loading for investigating dynamic response of human tympanic membrane exposed to blasts," *ASME J of Medical Diagnostics*, *8*(4), p. 041101, 2025.
11. Alipanahi, A., Oliveira Luiz, J., Cheng, J. T., Rosowski, J. J., and Furlong-Vazquez, C., "Blast production by a shock tube for use in studies of exposure of the tympanic membrane to high-intensity sounds," *Society for Experimental Mechanics Annual Conference and Exposition*, Orlando, FL, June 5–8, pp. 1–8, 2023.
12. Oliveira Luiz, J., Alipanahi, A., Cheng, J., Rosowski, J., and Furlong, C., "High-speed schlieren imaging of shock waves for the study of tympanic membrane's response," *Proc. ASME Int. Des. Eng. Tech. Conf. Comput. Inf. Eng. Conf.*, 12, V012T12A025, Boston, MA, Aug. 20–23, ASME, 2023.
13. Oliveira Luiz, J., Alipanahi, A., Furlong, C., Rosowski, J., and Cheng, J., "A parametric study using 3d-printed samples simulating human eardrums to investigate the middle ear's responses to high acoustical events," *Society for Experimental Mechanics Annual Conference and Exposition*, Vancouver, WA, June 3–6, 2024.
14. International Digital Image Correlation Society, Jones, E. M. C., and Iadicola, M. A., "A good practices guide for digital image correlation, international digital image correlation society," International Digital Image Correlation Society, Portland, OR, accessed Jan. 4, 2024, https://doi.org/10.32720/idics/gpg.ed1/print.Format, 2018.
15. Correlated Solutions, "Vic-3D 8 manual and testing guide," Correlated Solutions, Irmo, SC, accessed Feb. 9, 2025, https://correlated.kayako.com/article/57-vic-3d-8-manual-and-testing-guide, 2021.